Yours affectionately
WTR Wheldon

CAMBRIDGE UNIVERSITY PRESS
Cambridge, New York, Melbourne, Madrid, Cape Town,
Singapore, São Paulo, Delhi, Tokyo, Mexico City

Cambridge University Press
The Edinburgh Building, Cambridge CB2 8RU, UK

Published in the United States of America by Cambridge University Press, New York

www.cambridge.org
Information on this title: www.cambridge.org/9781107601222

First published 1906
First paperback edition 2011

A catalogue record for this publication is available from the British Library

ISBN 978-1-107-60122-2 Paperback

WALTER FRANK RAPHAEL WELDON. 1860—1906.[*]

A MEMOIR.

I. *Apologia.*

IT is difficult to express adequately the great loss to science, the terrible blow to biometry, which results from the sudden death during the Easter vacation of the joint founder and co-editor of this journal. The difficulty of adequate expression is the greater, because so much of Weldon's influence and work were of a personal character, which only those who have enjoyed his close friendship can estimate, and which will only to some extent be understood should it ever be possible to publish his scientific correspondence. That correspondence is not only the most complete record of the development of the biometric conceptions, but the amplest witness to Weldon's width of knowledge, keenness of intellectual activity, and intense love of truth. It is marked by an extreme generosity to both friend and foe, which is not in the least incompatible with the use of frankly—perhaps it would be better to say playfully—strong language whenever the writer suspected unfair dealing, self-advertisement, or slipshod reasoning masquerading as science. Any form of publicity was very distasteful to Weldon; in particular he had a strong dislike for all forms of personal biography. The knowledge of this makes the writing of the present notice a peculiarly hard task. Yet Weldon's influence and activity must always be associated with the early history of biometry; if there be anything which can effectively aid younger workers in this field, it must be to realise that at least one man of marked ability and of the keenest scientific enthusiasm has devoted the most fertile years of his life to this new branch of science. Weldon's history is not written in a long series of published memoirs; much of his best work was unfinished at his death, and we can only trust that it will eventually be completed as the truest memorial to his life. But science, no less than theology or philosophy, is the field for personal influence, for the creation of enthusiasm, and for the establishment of ideals of self-discipline and self-development. No man becomes great in science from the mere force of intellect, unguided and unaccompanied by what really amounts to moral force. Behind the intellectual capacity

[*] I have gratefully to acknowledge much aid from Mr A. E. Shipley in the preparation of certain parts of this memoir. K. P.

W. Memoir

1

there is the devotion to truth, the deep sympathy with nature, and the determination to sacrifice all minor matters to one great end.   What after all helps us is not that "he settled *Hoti's* business"...

> "Properly based *Oun*—
> Gave us the doctrine of the enclitic *De*,"

but that the Grammarian had the strength of will which enabled him "not to Live but Know."

If there is to be a constant stream of men, who serve science from love as men in great religious epochs have served the Church, then we must have scientific ideals of character, and these do involve some knowledge of personal life and development.   It is the abuse of the personal so prevalent in modern life, the mere satisfaction of a passing curiosity, which we have to condemn.   But the personal which enables us to see the force of character behind the merely intellectual, is of value, because it moulds our working ideals.   We see the environment—imposed and self-created—which favours scientific development, and we can with accumulating experience balance environment against heritage in the production of the highest type of scientific mind.   From the standpoint that no man works effectively without a creed of life, that for width of character and healthy development there must ever be a proper balance of the emotional and the intellectual, it would be a distinct loss if the personal were removed from what we know of the lives of Charles Darwin and James Clerk-Maxwell.   Science, like most forms of human activity, is occasionally liable to lose sight of its ultimate ends under a flood of controversy, the strugglings of personal ambition, or the fight for pecuniary rewards or less physical honours.   The safety of science lies in the inculcation of high ideals among its younger votaries.   A certain amount of purely human hero-worship is not to be condemned, and yet this is impossible without some knowledge of the personal.   Weldon himself was no more free from hero-worship than the best of his contemporaries.   Of the men whose influence tended most to mould his life and career—F. M. Balfour, T. H. Huxley, Francis Galton—the personal side was not the smaller element.   There was enthusiasm, hero-worship in its best sense, unregarding self-sacrifice in the defence of the man who had become for Weldon not only an ideal thinker, but an ideal character.   In the defence of hero or friend, Weldon belonged to a past age, he was out with his rapier, before considering the cause; it was enough for him to know that one he loved or admired was attacked. A criticism of Huxley was to the end inadmissible; if at any point apparently correct, this appearance of correctness was due solely to the inadequate manner in which the facts of his life had been reported by biographers,—the class who pandered to the public love of the petty.   It was in this spirit that Weldon received with delight the request to write for the *Dictionary of National Biography,* a scientific appreciation of Huxley's work.   From Weldon's standpoint that appreciation should have formed the "Life."   It is a fine piece of work and it was a labour of love, but those who have ever watched the younger man with the old, will know that the Huxley of the appreciation was not all that

Huxley meant to Weldon; the feeling of affectionate reverence did not spring from intellectual appreciation. It had far more its source in the influence of a strong character on a sympathetic character. And when we turn to Weldon himself, his relation to his friends and pupils was not purely that of a keen strong intellect; his best and greatest influence arose from the strength of character, that subtle combination of force and tenderness, which led from respect for the master, to keenest affection for the man.

If then we are to realise his life, it cannot be by a strict adherence to an appreciation of his published work. Some account of his stock, his early environment, and his temperament becomes needful, and the value of such an account lies in the help with which any life spent in single-eyed devotion to the pursuit of truth provides us, when we have ourselves to form our creed of life, and to grasp that science is something more than one of the many avenues to a competency. It must be in this spirit, therefore, that Weldon's dislike to the biographical is in a certain sense, not forgotten, but frankly disregarded in these pages.

II. *Stock and Boyhood.*

It would be impossible in a journal like *Biometrika*, devoted to the consideration of the effects of inheritance and environment, to pass by the striking resemblance of Raphael Weldon to his father Walter Weldon. The facts of Walter Weldon's life are given in the *Dictionary of National Biography*. It appears to have been a resemblance not only in intellectual bent, but also in many respects in emotional character. Raphael Weldon's paternal grandparents Reuben Weldon and his wife Esther Fowke, belonged to the manufacturing middle class. Their son Walter Weldon was born at Loughborough, October 31, 1832. Of his childhood we know little, he was as reticent as his son about both his childhood and his home surroundings; there is reason to suppose they were not wholly happy, and that shadows from these early years may have cast themselves not only over the father, but in a lesser extent have moulded the thought and life of the son. Walter Weldon married Anne Cotton at Belper, March 14, 1854, and shortly afterwards, leaving his father's business, came to London, starting as a journalist, writing for the *Dial* and *Morning Star*. Here he first made the acquaintance of William and Mary Howitt, who proved long and intimate friends of the family. From 1860 to 1864 he edited Weldon's *Register of Facts and Occurrences relating to Literature, the Sciences and the Arts*, and had as contributors a number of men afterwards well known in the world of letters. Thus while Walter Weldon's real name was to be made in science, his first interests were in literature and art. The steps by which Weldon regenerated the manganese peroxide used in the manufacture of chlorine, and the extensions he made of his chlorine process up to his death have been well described by Dr Ludwig Mond in his address in 1896 to the Chemical Section of the British Association. They brought Weldon comparative wealth, though nothing compared with the three-quarters of a million pounds his process saved this country annually. They also brought him scientific reputation; a vice-presidency of the Chemical Society, and in 1882 the fellowship of the Royal Society. But for our present purposes the main point is this: that Walter Weldon made his discovery while totally unacquainted with the methods of quantitative chemical analysis and possibly because of this ignorance. He was accustomed to attribute the discovery to a peculiar source, but those who knew well the immense facility of his son for closely observing phenomena out of his own field of research, and rapidly studying their interaction, always probing things, whether in the physical universe, or in mechanism, to their basis in simple laws of nature, will at once realise the source of the father's inspiration, and the heritage to the son*.

If Walter Weldon's discovery brought him wealth, he was generous to a fault. Like his son he appears to have scarcely known the value of money, except as a means of giving pleasure to his friends. His early death in September, 1885, two years after his son's marriage, cut off a career far from completed. But his life had been lived to the full, each instant crowded with physical, intellectual, or emotional activity. It is impossible to regard Walter Weldon's character without seeing whence Raphael Weldon drew much of his nature. The intense activity, the keen sympathy and generosity, the reticence, the creative power in many channels, the artistic appreciation†, were common to father and son. Nay, perhaps to give

---

* Raphael Weldon delighted during his many voyages in spending days in the engine-room; he made a study of the various types of engines, and his knowledge in this respect was not without service to the Marine Biological Association. He even studied the use of indicator diagrams. His first plan with a new bicycle was to take it part from part, so that he could fully understand its working and the nature of possible repairs. The microscope was not merely an instrument to work with, but a familiar illustration of optical laws, so that he knew at once how to modify each detail to suit special needs. Over and over again, talking over physical problems he would say: "Well, I don't know what you people think, but it has always seemed to me that "—and then would come some luminous suggestion or apt criticism of a proposed investigation in a field wholly outside the biological. A striking instance of this occurred only in the autumn of last year. Many friends had already gone to see the eclipse, most people were talking about it, and Weldon was left in sultry Oxford, fighting out a theory of determinantal inheritance. It was settled that a holiday should be taken, the determinants put on one side and a continuous photographic record made of the eclipse. Neither Weldon nor his colleague knew anything about sun-photography, and miserable were their first attempts. But gradually the objective, the telephoto lens and the focal shutter were worked out; a camera which had done yeoman service in photographing snail habitats became a wonderful structure, and a whole series of colour screens prepared from biological sources were tested and criticised. It was Weldon who obtained the first clean cut photograph showing sun spots clearly and admitting of definite enlargement. But what is more, each developmental stage of his sun camera had been thought out physically, and he knew why he took it. The trained physical astronomer would have found the stages already made, and *a posteriori* each would have been obvious, but this was the case of a biologist with insight into other fields and a striking power of making things work.

† An interesting illustration of the relationship is given in *Mary Howitt, an Autobiography*, 1889 (p. 184). The child Raphael, then 10 years, had gone with his father and the Howitts to visit the Wiertz Gallery at Brussels. William Howitt writes: "On our first entrance I was quite startled, I did not think I should at all like the paintings, they appeared so huge, so wild and so fantastic. But by degrees I began to see a great mind and purpose in them....... Little Raphael came and took my hand as we left the gallery, and said: 'Mr Howitt, I think Wiertz could not be a good man.' I asked him why. He answered, 'I think he could not be a good man, or he would not have painted some things there.' I told him he might naturally think so, but that a vast deal was to be allowed for his education. No doubt Wiertz thought all was right, and that many of his pictures contained

expression to a paradox, their volume of life was too great to be compatible with its normal length. There are men—not the least favoured of the Gods—who live so widely and so deeply, that they cannot live long. Discussions on the inheritance of longevity now come back to the memory, wherein Weldon referred to stocks of short-lived but intense life, and the personal experience and its moulding effect on character are now clear, where at the time the mind of the listener ran solely on a correlation coefficient.

In one respect Raphael Weldon differed widely from his father. Walter Weldon turned naturally to the mystical to satisfy his spiritual cravings; he was a Swedenborgian, and *ipso facto* a believer in intercourse with another world. Whether owing to a difference of training or of temperament, these things were to Raphael Weldon uncongenial. He was through the many years the present writer knew him, like his hero Huxley, a confirmed Agnostic. Sympathetic as every cultured mind must be with the great creations of religious faith; knowing more than many men of religious art—painting, sculpture, and music—he yet fully realised that these things had for him only emotional, no longer intellectual value*. It may be that the difference of training made this distinction between father and son, for the latter's mind was keenly alive to spiritual influences. A solitary fortnight with the beloved Dante was not solely pleasure; the re-perusal of the *Inferno* left its sombre influence on Weldon's thoughts for long after, testifying not only to its author's supremacy, but to the spiritual impressibility of the reader's nature.

It may be that the difference was due to heritage from the mother's side. Of Anne Cotton we know little, she died in 1881, when Raphael Weldon had just taken his degree. She appears to have exercised a rather stern discipline, which had greater influence on Raphael, than on his brother Dante. She was a devoted companion to Walter Weldon, and a resourceful helpmate in his early struggling days†. A daughter Clara born in 1855 died in 1861. Of his childhood Weldon rarely spoke. He was born in the Highgate district, and shortly after his birth his parents removed to a three-gabled house on the West Hill still standing. Here we get occasional peeps of a solitary child who would retire for hours under the dining-room table with his Shakespere, learning whole acts by heart. At six years old he appears in Mary Howitt's letters as staying at Claygate near Esher.

great and useful lessons. His father came up and added that when Raphael was older he would see those lessons more clearly than he could now."

The prophecy was fulfilled, in perhaps rather a different way. The little Raphael became a big Raphael who did not look to art "for great and useful lessons," and who refused to study Ibsen because undiscerning critics made current the idea that his art was subservient to inculcating a lesson.

* The "fulness of life" admitted, nay demanded, many a visit to cathedral service, especially in Italy. Even a study of Gregorian music was entered upon, and the writer recollects many a summer's afternoon spent in visiting the churches of Oxfordshire and Berkshire,—the cycle ride, the keen eye on surrounding nature, not only from the standpoint of the biologist, but of the artist; then the break to the religious past, the "biometric tea" at the village inn; the return journey towards evening and the discussion which touched many things, from *Draba verna* to the Norsemen in Sicily. The "volume of life" was there, as it was in the midnight talks in Wimpole Street or in the discussions in the study at Merton Lea.

† See *R. S. Proc.* Vol. XLVI. "Obituary Notice of Walter Weldon," p. XIX. *et seq.*

" We find little Raphael Weldon one of the best of children. Secker is mowing the grass at this moment, and he harnessed like a pony is drawing the machine. The Pater calls him ' Young Meritorious.' " And again :

"[Agnes] and Raphael are the best of friends, and their ringing laughter comes to us in the garden through the open window, as they sit in the dining-room painting the Stars and Stripes and the Union Jack for each other's amusement.... Agnes is a little free-spoken American full of fun and *dash*. Raphael more silent and contemplative. They sit painting pictures together for hours at a time. I feel quite proud of them both."*

In 1870 comes the flying visit to Brussels ; in 1872 a still more memorable first visit to Paris, where the destruction caused by the Commune to the Tuileries and other buildings much impressed the boy. The Weldons had moved meanwhile to The Cedars, Putney, and shortly afterwards went to the Abbey Lodge, Merton, near Wimbledon. The visits and the changes give one the impression of a rather broken education. We have no record of what school Raphael Weldon attended, if any, at Highgate. At Putney he had as tutor a neighbouring clergyman. In 1873 he was sent to a boarding-school at Caversham, and from this time onwards the educational career is more definite.

Even before 1870, however, we find in the boy the father of the man. His great pleasure was to organise lectures for his children friends, and the adult population, if it could be procured. The seats were formally arranged, tickets provided, and the boy would discourse on slug or beetle procured in the garden, observation and the scanty literature available providing the material. According to a surviving auditor the lectures were carefully prepared and good so far as they extended.

Of the school at Caversham we have some detailed information. Mr W. Watson, its headmaster, had been a private 'coach' in London to University College students. In 1865 he opened a school at Reading, which was transferred to the hill out of Caversham in 1873†. Mr Watson's daughter Ellen Watson had a brief but brilliant career as a mathematician and pupil of W. K. Clifford's. Her life has been written by Miss Buckland. It is possible that she first stirred Weldon's mathematical tastes, as he spoke with admiration of her powers ; she does not, however, appear to have taught in the school. The pupils were chiefly sons of Nonconformists of some eminence. Among the earlier scholars were Viriamu Jones, Alfred Martin, and E. B. Poulton, and among the later pupils Owen Seaman, F. W. Andrewes, P. Jacomb-Hood, and W. F. R. Weldon ; names afterwards distinguished in literature, science, or art. The headmaster appears to have been a clever man of wide knowledge and sympathy, but there was little to specially encourage biological tastes in the school. It is reported of one under-

* *Loc. cit.* p. 162 *et seq.*

† As an illustration of Weldon's reticence I may state that we had passed this house several times together, before he mentioned it as his old school.

Raphael Weldon Aged 10.

master that he protested against the study of insects, asking: " How do you think that such pursuits will put a leg of mutton on your table?" and the ability that proceeded from the school has been attributed by one of its former pupils to the special class from which it drew its chief material.

III. *Lehrjahre.*

Weldon did not remain fully three years at this school. It was followed by some months of private study and he matriculated at 16 (1876) in the University of London. In October of '76 we find him at University College taking classes in Greek, English, Latin, and French, with two courses of pure mathematics. In the summer term of 1877 physics and applied mechanics were studied. During this whole session he also attended Daniel Oliver's general lectures on botany and Ray Lankester's on zoology. He used to come up to town for Oliver's 8 o'clock lectures, getting his breakfast at a bun-house on the way*. Of his education at University College he especially praised in after years Olaus Henrici's lectures on mathematics. They were he held most excellent, and he considered Henrici the first born teacher under whom he came. Later in the Christmas vacation of 1879, after he had gone up to Cambridge, he researched for some weeks under Ray Lankester, who set him to work out the structure of the gills of the mollusc *Trigonia.* This completes Weldon's relations as a student to University College.

The difficulty of access, or possibly Walter Weldon's strong views, led Raphael Weldon in the autumn of 1877 to transfer himself to King's College. Here he stayed for two terms attending classes in chemistry, mathematics, physics, and mechanics, beside the zoology course of A. H. Garrod and the biology of G. F. Yeo. Divinity under Barry, at that time I believe compulsory, was also taken. At this time Weldon had the medical profession in view. He was only entered on the Register of Medical Students on July 0, 1878, but there can be no question that his course on the whole was directed towards the Preliminary Scientific Examination of the London M.B. This examination he took in December, 1878, after he had gone up to Cambridge; he was coached for it by T. W. Bridge, now Professor of Zoology in Birmingham, but he had already completed the bulk of the work in his London courses. With the Preliminary Scientific, Weldon's relation to London ceased. His student career there was not of quite two years' duration and it dealt with a variety of subjects, dictated as much by Weldon's catholic tastes, as by the discursiveness of the London examination schedule. But in his case, as in that of others, the grounding he received in physics and mathematics became a valuable asset, and the taste for languages, afterwards so emphasised, was to some extent trained and coordinated with literary knowledge. Yet Weldon's earlier instinct to study biology was not substantially modified either by the choice of medicine as a profession or by the diversity of his London studies. In 1877 he attended the Plymouth Meeting of the British Association, and there he was generally to be found in Section D.

---

* Weldon states in his applications for the Jodrell and the Linacre Chairs that he commenced the study of zoology under Lankester in 1877.

The presence of a life-long friend, who had already gone to Cambridge, was at least one of the causes which led to Weldon's entering himself as a bye-term student at Cambridge, and probably his choice of St John's College was due to Garrod's influence. He was admitted on April 6, 1878, as a pupil of S. Parkinson's. In the record his father is given as a "Journalist," although the chlorine process had now become a success, and his reference is to the Professor of Mathematics at King's College, then W. H. Drew *.

At Cambridge Weldon soon found his work more specialised and he rapidly came under new and marked influences. His first May term and Christmas term were devoted to his preparation for Little-Go and the London Preliminary Scientific. For the classical part of the former he seems to have worked by himself. After these examinations were over reading for the Tripos was begun and, under the influence of Balfour, Weldon's thoughts turned more and more to zoology, and the medical profession became less and less attractive. During the years 1879 and 1880 Weldon worked steadily for his Tripos; in the first year he was given an exhibition at St John's, and almost the only break in his work was the York Meeting of the British Association. In the second year a little original investigation on beetles was started; in May he took, for a month, Adam Sedgwick's place and demonstrated for Balfour. Overwork led to a serious breakdown, and resulted in insomnia and other ills, which occasionally troubled him again in later life. At the annual British Association holiday, this year in Swansea, Weldon saw for the first time Francis Galton, but an actual friendship was not begun till some years later.

The Tripos work was continued in spite of ill-health, till the Easter of 1881, when Weldon was unable to enter for the college scholarship examinations. By the influence of Francis Balfour, however, Weldon's real ability was recognised and a scholarship was awarded to him. A three months' holiday had become necessary, and Weldon went to the south of France, returning only shortly before his Tripos examination. At the very start of this, in itself all-sufficient, mental strain, Dante Weldon, who had joined Peterhouse, died suddenly of apoplexy. It says much for Weldon's self-control that the terrible shock of his brother's death, though it greatly affected him, did not interfere with his place in the first class of the Natural Sciences Tripos. The distress he had felt at his brother's death was redoubled a few weeks later when his mother passed away. She had never recovered from the blow resulting from the tragic death of her younger son. Of these things Weldon did not speak, but they undoubtedly influenced immensely his deeply emotional nature. Balfour's untimely death in the following year, and the early death of Weldon's father a few years later, left also their indelible impresses, a certain tinge of melancholy, a doubt whether he too would live to finish his work, and a tendency to take the joy and fulness of life while it was there. Few who saw the almost boyish delight in work and in play, the energy which spent itself for hours at a problem, or cycled eighty or a hundred

---

* There are errors in the entries in the Register, Weldon's mother's maiden name is erroneously given as Chester, not Cotton. Weldon was actually born at Suffolk Villa, Highgate.

miles in the day, the activity in debate, the vigour in lecture, the flow of thought and talk to the midnight hours, realised that the man was not of iron physique, and had indeed but small reserves of strength. To see Weldon keen over a piece of work was to believe him robust and ready for any fray; but looking back on the past one can see what each piece of work cost him, and the strain on a highly nervous temperament began in even those early Cambridge days.

IV. *Wanderjahre.*

With the Tripos Weldon's *Lehrjahre* closed and, as his nature directed, the *Wanderjahre* began without any interval of rest. Immediately after his Tripos, Weldon started for Naples to work at the Zoological Station. We have seen that at Cambridge he had been a pupil of F. M. Balfour's, whose death from an Alpine accident in the July of 1882 was the greatest loss British zoology had sustained for years. The charm of Balfour's personality had aroused the affection of all who attended his classes, and had awakened a keen desire to follow, even if but a long way behind, in his footsteps. In those days the stimulus given by Darwin's writings to morphological and embryological researches was still the dominating factor amongst zoologists, and Weldon threw himself at first with ardour into the effort to advance our knowledge by morphological methods. In Naples he began his first published work, a "Note on the early Development of *Lacerta muralis*"(1), but at the same time did much miscellaneous work on marine organisms. The lizard paper was finished in the winter at Cambridge, Weldon gratefully acknowledging the help of Adam Sedgwick, in whose laboratory he was then working. Anticipation in the publication of some of the results by C. K. Hoffmann, who had been working at the same points, caused a not unnatural disappointment.

In September Weldon was back in England at the Southampton meeting of the British Association. Here Adam Sedgwick, who had succeeded to the teaching work of Francis Balfour, invited Weldon to demonstrate for him. Thus the winter found Weldon in Cambridge again, and from Sedgwick's laboratory was issued the next piece of work: "On the Head-Kidney of *Bdellostoma*, with a suggestion as to the Origin of the Suprarenal Bodies" (2). Weldon hoped to show that "at all events in Reptiles and Mammals, the connection between the Wolffian body and the suprarenal is much more intimate than has generally been supposed," and he followed the matter up in the next year by publishing his paper "On the Suprarenal Bodies of Vertebrates" (3).

Meanwhile a great change had come over Weldon's personal life. On March 14, 1883, the anniversary of his parents' wedding-day, he was married to Miss Florence Tebb, the eldest daughter of William Tebb, now of Rede Hall, Burstow, Surrey, which formerly, after he left Merton, had been the house of Walter Weldon. The Weldons and Tebbs had been intimate friends for many years, and Miss Tebb had been at Girton while Raphael Weldon was at St John's. At Cambridge the new Statutes had just come into force, marriage was the order of the day, and houses were even difficult to procure. The Weldons on their return from a tour in France took Henry Fawcett's furnished house and settled down in Cambridge for the May term. Raphael Weldon still had his scholarship, and he was demonstrating for Sedgwick. He was now compelled to undertake "coaching,"—work which he gave up as soon as his means would allow of it, for his whole heart was then as afterwards in research. Still this coaching work brought him in touch with many men who afterwards distinguished themselves in biological or other fields.

After the death, on the 14th January, 1883, of Forbes, a fellow Johnian, Weldon for four months—June 15 to October 15—acted as *locum tenens* for the Prosector at the Zoological Gardens, London, and during that time he read the following papers before the Zoological Society: "On some points in the Anatomy of *Phoenicopterus* and its Allies" (4); a "Note on the Placentation of *Tetraceros quadricornis*" (5), and "Notes on *Callithrix gigot*" (6). Weldon did not succeed Forbes—it was rumoured that some of the electors doubted the fitness of his physique for the work and considered that the post was not without danger. But the temporary work into which he threw his usual energy gave him increased insight into vertebrate anatomy and had the further advantage of making him personally known to the active workers in zoology of the metropolis.

In the following year (1884) the paper above referred to on the development of the suprarenal bodies was published in the *R. S. Proceedings*. Weldon was now demonstrating in comparative anatomy at Cambridge, and the holidays were devoted to collection. At Easter Banyuls was visited, and the summer vacation found Weldon in Naples again for three months preparing his fellowship dissertation. In Naples the cholera had broken out, and the Weldons experienced not only difficulty in getting the precious dissertation back to England, but in returning themselves. This was done by an Orient liner, the last allowed to call. Thus began the long series of holidays in Italy with the sea passage to or fro. The summer heat of Naples seemed to suit Weldon, and he could work and think under circumstances which only allow mere existence to an ordinary Englishman. On returning to Cambridge, Weldon was elected to a fellowship at St John's College on November 3rd, and was shortly afterwards appointed University Lecturer in Invertebrate Morphology. About this time the Weldons took a permanent home at No. 14, Brookside, which soon became a centre for Cambridge workers in biology*. The Weldons' home, whether in Cambridge, London, or Oxford, was always a centre, where not only the right people met, but whence actual profit came by the right people interchanging ideas and planning work.

On his return to Cambridge in November 1884 Weldon had taken up again his invertebrate work. His next memoir "On *Dinophilus gigas*" (7) dealt with the anatomy and affinities of *Dinophilus*, at that time a very little known Annelid. A. E. Shipley had been fortunate enough to collect a number of these minute worms at Mounts Bay, Penzance, and had handed them over to Weldon for

---

* "The house in Brookside in which he lived after his marriage until he left Cambridge was a delightful and hospitable centre, where all sorts of subjects were discussed, attacked and defended until all sorts of hours in the morning." A. E. S.

description. The latter gave a full account of their anatomy and added a careful discussion on the affinities of the genus, expressing his belief that while it is "related on the one hand to *Archiannelida*, it retains on the other many features characteristic of the ancestor common to those groups (especially Chaetopods, Gephyreans, Mollusca, Rotifers and Crustacea) which possess a more or less modified trochosphere larva."

The next few years of Weldon's life were—if it be possible to make any comparative where all were intense—more active than ever. He had now given up coaching, and as he only needed to be in Cambridge two terms of the year, travel and research could occupy the time from the beginning of June to January. On May 8, 1885, Weldon gave his first Friday evening lecture at the Royal Institution on "Adaptation to surroundings as a factor in Animal Development." No report is published in the *Proceedings* of this lecture, but there are those who still remember the impression caused by the youthful lecturer of 25 years of age. And here may be a fitting place to say something of Weldon's lecturing power. There are two distinct sides to lecture work; the instruction of small or large classes of students and the public oration. Success in the one field does not necessarily connote success in the other. In the former case the eye must be kept on the average student, the lecturer must realise what the individual auditor is feeling, he must expand his exposition or must contract it to meet the carefully observed needs of his audience, for he knows that he can take up the subject again on the next occasion exactly where he has left off. In this form of lecturing Weldon was an adept, it brought out all his force and enthusiasm as a teacher. As a writer in the *Times* (April 18, 1906), says:

"Seldom is it given to a man to teach as Weldon taught. He lectured almost as one inspired. His extreme earnestness was only equalled by his lucidity. He awoke enthusiasm even in the dullest, and he had the divine gift of compelling interest."

In public lecturing on the other hand, with a time limit and an unknown audience, the personal touch with individuals is impossible. There is no time to elaborate points, the whole matter must be *a priori* fitted to the time, and if the audience is not grasping an idea, then the lecturer must put both explanation and disappointment on one side; he must make his audience jump gladly, and trust to better luck in his exposition of the next stage of his thesis. Shortly, he must feel his audience with him *as a whole* and pay no regard to the individual.

Weldon's own intense thoroughness made him only too conscious when a portion of his audience were not following him; his highly nervous temperament made it a necessity that he should have a sympathetic grip on the individual. This made for success in his lectures to students; but it brought also a factor of uncertainty into his public lectures. The most carefully prepared discourse, and no man gave more time and energy than Weldon to preparation*, might be

* Drafts and re-drafts were written, elaborate diagrams painted, or lantern slides made and coloured by Weldon himself.

spoiled by Weldon's consciousness that certain members of the audience were not following him. He would then turn his exposition into explanation of minor points, so that the lecture would not be completed, or he would settle down to speak to the few he realised to be following him, and neglect the audience as a whole. If a portion of his audience were hostile or actively unsympathetic, this always prevented Weldon from reaching his best; it formed a strain on the lecturer's nervous temperament, which could only be realised by those of like fibre, and in some cases left its permanent mark. Thus it came about that the success of a public lecture by Weldon could not *a priori* be measured; it depended far too much on the audience. Individual lectures at the British Association, the London Institution, at University College or elsewhere were brilliant achievements, but at the same places on other occasions, Weldon was not so successful, for no man was ever more responsive to immediate environment than he was. To do his best and to be at his best he needed essentially a sympathetic environment. Weldon has been spoken of as an eager, ready and dramatic debater, keen to see a weak joint in his opponent's armour and quick in putting his own case with telling effectiveness. This is undoubtedly true, but it needs the qualification that this intellectual readiness when in full action meant a high pressure; it was a strain the less oft repeated the better. A torpedo-boat destroyer is associated with a 26 knot speed, and such speed differentiates it from other vessels of war; but the less it is run at this rate, the longer undoubtedly it will last. Controversy was not an atmosphere in which Weldon rejoiced*; it came to him because he felt bound to criticise what he held to be error, because he must defend a friend, but it was—running the destroyer at 26 knots!

This digression may be justified on the ground that we have reached the period when Weldon began to exercise a personal influence over his students at Cambridge, and the sources of that influence are to be found first in the lecture-room and then in strong personal sympathy. In the lecture-room he always impressed his hearers with the importance of his topic. You could not listen to him lecturing on a flame-cell or on the variations in the carapace of *Pandalus annulicornis* without sharing his intense conviction of the importance of the matter in hand. He aroused a consciousness in his students that things were worth studying for their own sake, apart from their examination value.

The summer months of 1885 were spent in Guernsey, and the death of Walter Weldon in the September of this year kept Raphael Weldon at other things than research. Christmas, however, found the Weldons at Rome. The Lent and May terms (1886) were spent in Cambridge as usual. In June came a visit to the south of France on Chlorine business, but in July came freedom, the crossing to America and the visit to the Bahamas in August to collect. From his

---

* Actual experimental work which upset another man's views, Weldon declined to publish. " Yes, I know he is wrong, but I don't want merely to controvert him, I want to get at the truth of these things for myself." And when he had satisfied himself he would pass on to a new point of investigation and never publish at all.

headquarters in the Bahamas Weldon went with two friends to North Bimini in the Gulf stream and enjoyed immensely his first experience of tropical or at least semi-tropical seas. He made considerable collections, but his published results were confined to "*Haplodiscus piger*; a new Pelagic organism from the Bahamas" (8), and a "Preliminary Note on a *Balanoglossus* Larva from the Bahamas" (9). Haplodiscus was netted near the Island of New Providence. It is a member of the Acoela, the most simplified of the class Turbellaria, and for some time Weldon's account was one of the most complete we had of any member of this group. Working at the *Balanoglossus* material in 1887, Weldon found that his results differed from those reached by Professor Sprengel. He accordingly went to Giessen at Easter,—his second visit to Germany, the first having been at Christmas, 1886—and finally handed over to Professor Sprengel the whole of the *Balanoglossus* material he had collected in the Bahamas. During the Lent and May terms Weldon came up from Cambridge and gave a course on Economic Entomology to the forestry students at the Royal Engineering College, Cooper's Hill. The summer and autumn of this year involved a meeting of the British Association at Manchester, a visit on Chlorine business to France, and later, collecting and working in Guernsey. The Christmas was spent at Plymouth.

In 1888 the buildings of the Marine Biological Laboratory in Plymouth were nearly completed, and the visits to Plymouth now replaced those to Guernsey. To the Marine Biological Association Weldon gave both time and sympathy during the rest of his life. His annual visits of inspection to Lowestoft during the last few years were always a great pleasure to him, and he was preparing for and talking of this year's visit only a few days before his death. Lent and May terms, 1888, were spent as usual in Cambridge, but June to December were given up to Plymouth, with a brief Christmas holiday in Munich. And here we must note the beginning of a new phase in Weldon's ideas. His thoughts were distinctly turning from morphology to problems in variation and correlation. He has left on record the nature of the problems he was proposing to himself at this time and they are summed up as follows:

(1) The establishment of a new set of adult characters leading to the evolution of a new family has always been accompanied by the evolution of a new set of larval characters leading to the formation of a larval type peculiar to the newly established family; the two sets of characters having as yet no demonstrable connection one with the other.

(2) The evolution of the adult and that of the larval characters peculiar to a group advance *pari passu* one with the other, so that a given degree of specialisation of adult characters on the part of a given species implies the possession of a larva having a corresponding degree of specialisation and *vice versa*.

The next year was to place in Weldon's hands a book—Francis Galton's *Natural Inheritance*, by which one avenue to the solution of such problems, one quantitative method of attacking organic correlation, was opened out to Weldon;

and from this book as source spring two of the friendships and the whole of the biometric movement, which so changed the course of his life and work. In 1889, the year of the issue of this book, another change also came. Weldon found that his dredging and collecting work separated him from his books for half his time. Accordingly, he applied for a year's leave from Cambridge, and the Weldons settled down in a house of their own at Plymouth. This period of hard work lasted through 1890, and was broken only by flying visits to Dresden in September and at Christmas, 1889, and an autumn visit in 1890 to Chartres and Bourges. The intellectual development and the experience and knowledge gained in this period were far more important than the mere published work would indicate. In 1889, Weldon investigated the nature of the curious enlargement of the bladder associated with the green, or excretory, glands in certain Decapod Crustacea, and published in October of the same year his paper on "The Coelom and Nephridia of *Palaemon servatus*" (10). The result of his investigation was to confirm "the comparison so often made (by Claus, Grobben, and others) between the glomerulus of the vertebrate kidney and the end-sac of the Crustacean green gland." A little later, June 1891, Weldon published the results of more extended researches in this field in what proved to be his last strictly morphological paper. It was entitled: "The Renal Organs of certain Decapod Crustacea" (11). In this he showed that in many Decapods spacious nephro-peritoneal sacs "should be regarded rather as enlarged portions of the tubular system...than as persistent remnants of a 'coelomic' body cavity into which the tubular nephridia open."

One further paper of a year later may be best referred to here, Weldon's only piece of work on invertebrate embryology, "The Formation of the Germ Layers in *Crangon vulgaris*" (12). This contains a clear account of the early stages of segmentation and the building up of the layers of the shrimp, illustrated by excellent figures. And here it may be mentioned that Weldon's power with the pencil was not that of the mere draughtsman, accurate in detail, but too often lifeless. Weldon was an artist by instinct, and he had the keenest pleasure in drawing for its own sake. His brilliant blackboard drawings will be remembered by all his students; some correspondents will remember elaborately beautiful sketches sent merely to illustrate a passing question, where a rough diagram would have sufficed; a delicately pencilled shell to please a child friend; carefully copied architectural details to gratify himself and made to be destroyed;—all signs of a real artistic power of creation. And the sense he enjoyed in himself, he appreciated in others. Nothing refreshed him so much as a visit to the National Gallery, or to a lesser extent the sight of more modern art. Weldon, smiling before one of his own pictures, unconscious of his environment, was good to behold, and made one realise how for him pictures were still differentiated from furniture. In the last two years of his life, when he had become an ardent photographer, the artistic feeling came to play a prominent part as the difficulties of the craft were one by one mastered.

(a) "L'Apparition : Le Café Orleans."

(b) H. Hortensis, from a letter.

December, 1890, closed the Cambridge work* and concluded the *Wanderjahre.* Weldon now succeeded Ray Lankester in the Jodrell Professorship at University College. In June he had been elected a Fellow of the Royal Society largely on the basis of his first two biometric papers, which will be considered more in detail in the next section.

It will be seen that the years between Weldon's degree and his first professoriate were years of intense activity. He was teaching many things, studying many things, planning many things. His travels perfected his linguistic powers, and his fluency in French, Italian and German was soon remarkable. But while this added immensely to his delight in travel, it opened to him also those stores of literature, which appealed so strongly to his artistic temperament. From the mediaeval epics to Balzac he was equally at home in French literature; and the Italian historians were read and carefully abstracted, that he might understand Dante without the aid of a commentator, and appreciate Italian towns without the help of a guide-book. In German he had a less wide knowledge of the earlier literature and history, but he spoke the language with an accent and correctness remarkable in an Englishman. In later years he had commenced the study of Spanish, the Romance tongues and literatures being always more sympathetic to him than the Scandinavian or Teutonic. His remarkable thoroughness in science reappeared as a form of scholarly instinct when he approached history and literature, and the present writer remembers Weldon's keen pleasure and exactitude in following up more than one historical enquiry. His delight in knowing spread far beyond the limits of natural science.

V. *London and the First Professoriate,* 1891–1899.

A word must here be said as to the transition which took place during the *Wanderjahre* in Weldon's ideas. He had started, as most of the younger men of that day, with an intense enthusiasm for the Darwinian theory of evolution; it threw open to him, as to them, a wholly new view of life with its possibility of seeing things as a connected whole. Weldon realised to the full that the great scheme of Darwin was only a working hypothesis, and that it was left to his disciples to complete the proofs, of which the master had only sketched the

---

* A note may be added as to the general influence of Weldon at Cambridge. At the time Weldon began lecturing there were a considerable number of students largely attracted to Cambridge by Balfour's fame and remaining there to mourn his loss. Mr W. Bateson of St John's, Dr Harmer of King's, Professor Sherrington of Caius, Professors D'Arcy Thompson and J. Reynolds Green of Trinity, Professor Adami and Mr A. E. Shipley of Christ's, graduated in 1883 and 1884, and all, to some extent, came under his influence. For six years (1884–1890) he gave advanced lectures to the candidates for Part II of the Natural Sciences Tripos. During these few years the number of men in his class who have since done much to advance science was considerable. The following is by no means a complete list. Among botanists, F. W. Oliver, C. A. Barber, W. B. Bottomley; among geologists, T. T. Groom, P. Lake, S. H. Reynolds, H. Kynaston and H. Woods; among physiologists, pathologists and medical men, A. E. Durham, H. E. Durham, J. S. Edkins, W. B. Hardy, A. P. Beddard, E. H. Hankin, H. Head; and among zoologists, H. Bury, G. P. Bidder, W. F. H. Blandford, R. Assheton, F. V. Theobald, T. H. Riches, E. W. MacBride, H. H. Brindley, A. T. Masterman, C. Warburton, and Malcolm Laurie.

outline. Naturally he turned first to those methods of proof, morphological and embryological, which were being pursued by the biological leaders of the period, and it was only with time that he came to the conclusion that no great progress could be attained by the old methods. We have already seen that even before the appearance of *Natural Inheritance*, Weldon's thoughts were turning on the distribution of variations and the correlation of organic characters. He was being led in the direction of statistical inquiry. The full expression of his ideas is well given in the first part of the " Editorial " with which *Biometrika** started :

"The starting point of Darwin's theory of evolution is precisely the existence of those differences between individual members of a race or species which morphologists for the most part rightly neglect. The first condition necessary, in order that any process of Natural Selection may begin among a race, or species, is the existence of differences among its members ; and the first step in an enquiry into the possible effect of a selective process upon any character of a race must be an estimate of the frequency with which individuals, exhibiting any degree of abnormality with respect to that character, occur. The unit, with which such an enquiry must deal, is not an individual but a race, or a statistically representative sample of a race ; and the result must take the form of a numerical statement, showing the relative frequency with which various kinds of individuals composing the race occur."

It was Francis Galton's *Natural Inheritance* that first indicated to Weldon the manner in which the frequency of deviations from the type could be measured. A mere catalogue of exceptional deviations seemed to him of little value for the study of Natural Selection. But this description of frequency was only the first stage. How did selection leave the distribution? and How was the intensity of selection to be measured? naturally arose as the next problems. These problems led at once to the even greater question of the influence of selection on correlation. What is the relation between organs in the same individual, and how is this changed, if at all, by the differentiation of species, or at least by the establishment of local races? Nor could the problem of evolution be complete without ascertaining the manner in which deviations were inherited. The modern biometric methods of discussing these problems, if very far from fully developed, were at least suggested in Galton's great work, and that book came as a revelation not only to Weldon, but to others who were preparing to work on similar lines†.

In Plymouth, 1890, Weldon started his elaborate measurements on the Decapod Crustacea and soon succeeded in showing that the distribution of variations was closely like that which Quetelet and Galton had found in the case of man. So far as the present author is aware, the paper "The Variations occurring in certain Decapod Crustacea I. *Crangon vulgaris*" (13) was the first to apply the methods of Galton to other zoological types than man‡. In this paper Weldon shows that different measurements made on several local races of shrimps give frequency distributions closely following the normal or Gaussian law. In his next paper,

---

* Vol. I. p. 1.

† The present writer's first lecture on inheritance was given on March 11, 1889, and consisted of an exposition and amplification of Galton's theory.

‡ Galton had dealt with the weights of sweet pea seeds, Merrifield with the sizes of moths, but they had not published fitted frequency distributions.

"On certain correlated Variations in *Crangon vulgaris*" (14), Weldon calculated the first coefficients of organic correlation, i.e. the numerical measures of the degree of interrelation between two organs or characters in the same individual. It is quite true that the complete modern methods were not adopted in either of these papers, but we have for the first time organic correlation coefficients—although not yet called by that name—tabled for four local races. These two papers are epoch-making in the history of the science, afterwards called biometry.

It is right to state that Weldon's mathematical knowledge at this period was far more limited than it afterwards became. The first paper was sent to Francis Galton as referee, and was the commencement of a life-long friendship between the two men. With Galton's aid the statistical treatment was remodelled, and considerable modifications made in the conclusions. But the credit of making the vast system of measurements, of carrying out the necessary calculations (now with the aid of his wife, who was for years to assist in this part of the work), of seeing *a priori* the bearing of his results on the great problems of evolution, must be given to Weldon. Nor must we forget the rich suggestiveness of these papers. Weldon was on the look-out for a numerical measure of species. He was seeking for something constant for all local races, and although his suggestion that the correlation coefficient was a constant for local races has not been substantiated— the "selection constant," the quantity uninfluenced by racial differentiation, being of a much more complex nature—yet his suggestion directly led up to the investigation of correlation in man, animals and plants, and has given us immensely clearer ideas on the inter-relationship of organic characters. And Weldon realised this also :

"A large series of such specific constants would give an altogether new kind of knowledge of the physiological connexion between the various organs of animals ; while a study of those relations which remain constant through large groups of species would give an idea, attainable in no other way, of the functional correlations between various organs which have led to the establishment of the great sub-divisions of the animal kingdom*."

The defect in mathematical grasp, which Weldon had realised in his first paper, led him at once to seek to eliminate it. He sought first to 'enthuse' a mathematician with his project of demonstrating Darwinian evolution by statistical enquiries. A visit was paid to Cambridge with this end in view, but it did not lead to the required result. Weldon then set about increasing his mathematical knowledge by a thorough study of the great French writers on the calculus of probability. He did not turn to elementary text-books but with his characteristic thoroughness went to the fountain head. Turning over his papers now, it is astonishing to notice the completeness of his studies as evidenced by his notes and abstracts. He thus attained to a very great power of following mathematical reasoning and this power developed with the years. He never reached a high wrangler's readiness in applying analysis to the solution of new questions, possibly this requires years of training in problem papers; but he was able to follow and

* *R. S. Proc.* Vol. LI. p. 11.

criticise extremely complicated algebraical investigations, and to reproduce and often simplify them for the use of his own students. He had, however, a touch with observation and experiment rare in mathematicians. In problems of probability he would start experimentally and often reach results of great complexity by induction. Thus he was able to find out a number of problems relating to the correlation between a throw of n dice, and the result obtained when a re-throw of m out of the number n was made, and others relating to the mixture of n packs of cards and the throwing out of random portions*. In all these cases Weldon was illustrating by a game of chance a definite biological process.

From 1890 onwards, Weldon's knowledge, theoretical and experimental, of the theory of chance increased by bounds. Weldon and the present writer both lectured from 1 to 2, and the lunch table, between 12 and 1, was the scene of many a friendly battle, the time when problems were suggested, solutions brought, and even worked out on the back of the *menu* or by aid of pellets of bread. Weldon, always luminous, full of suggestions, teeming with vigour and apparent health, gave such an impression to the onlookers of the urgency and importance of his topic that he was rarely, if ever, reprimanded for talking 'shop.'

It is difficult now, after fifteen years of common work and continuous interchange of ideas, to distinguish where one or other idea had its source, but of this the writer feels sure, that his earliest contributions to biometry were the direct results of Weldon's suggestions and would never have been carried out without his inspiration and enthusiasm. Both were drawn independently by Galton's *Natural Inheritance* to these problems, but the papers on variation and correlation in shrimps—which in rough outline are types of all later biometry—were published before their friendship had begun.

Weldon's work at University College commenced in 1891. The house in Wimpole Street was taken and, if possible, life became more intense. Easter was spent at Chartres. In the summer came the annual visit to Plymouth, where work on crabs was now to replace that on shrimps. September gave some rest with a sea trip to Malta. In October came the college inaugural lecture for the session, Weldon taking as his subject the statistical treatment of variation. At Christmas there was a break for opera in Munich and Dresden. This year and the next were strenuous years in calculating. The Brunsviga was yet unknown to the youthful biometric school; the card system of correlating variables was still undeveloped, we trusted for multiplication to logarithms and Crelle, and computors trained to biometric work had to be created. The Weldons toiled away at masses of figures, doing all in duplicate. At Easter, 1892, they went to Malta and Naples, and the summer was spent over crab-measurements at the zoological station in the

---

* In the summer of 1905 a great deal of work was done by the present writer in conjunction with Weldon on mixtures of card packs, the main features of the work having been already outlined by Weldon. The results are summed up in a theory of determinantal inheritance which, it is hoped, will be eventually published.

latter city, and the first biometric crab paper "On certain correlated Variations in *Carcinus moenas* " (15) was issued in this year. In this paper Weldon confirms on the shore crab his results for the common shrimp. The distributions of characters are closely Gaussian with the exception of the relative frontal breadth, which Weldon considered dimorphic in Naples, a problem which led to the present writer's first paper in the *Contributions to the Mathematical Theory of Evolution.* It is right to say that Weldon had reached a moderately accurate solution by trial and error before he proposed the problem to his colleague. He does not refer to this fact in his memoir. As for shrimps the correlations again came out closely alike for the Plymouth and Naples races. Weldon was not dogmatic on the point; he considered the constancy as at least an "empirical working rule" and this it has certainly proved.

"The question whether this empirical rule is rigidly true will have to be determined by fuller investigation on larger samples ; but the value of a merely empirical expression for the relation between abnormality of one organ and that of another is very great. It cannot be too strongly urged that the problem of animal evolution is essentially a statistical problem : that before we can properly estimate the changes at present going on in a race or species we must know accurately (*a*) the percentage of animals which exhibit a given amount of abnormality with regard to a particular character ; (*b*) the degree of abnormality of other organs which accompanies a given abnormality of one ; (*c*) the difference between the death rate per cent. in animals of different degrees of abnormality with respect to any organ ; (*d*) the abnormality of offspring in terms of the abnormality of parents and *vice versâ*. These are all questions of arithmetic ; and when we know the numerical answers to these questions for a number of species we shall know the deviation and the rate of change in these species at the present day—a knowledge which is the only legitimate basis for speculations as to their past history, and future fate."

These concluding words were surely epoch-making; they formulated the fundamental principles of biometry. We may criticise the memoir in that the index measurements selected by Weldon overlooked the question of spurious correlation, or because the growth law of the indices had not been previously determined. But these are minor matters compared with the general ideas involved in the memoir. It is a paper which biometricians will always regard as a classic of their subject. It first formulated the view that the method of the Registrar-General is the method by which the fundamental problems of natural selection must be attacked, and that is the essential feature of biometry.

Besides biometry a new bond drew Weldon and the present writer together. Since 1884, a strong movement for the reform of the University of London had been in progress, association followed on association, royal commission on royal commission. Few people had distinct ideas of what they themselves wanted, scarcely any one had a notion of what a real university must connote. At University College, after severe crises, the teaching staff had won direct representation on the governing body, and was beginning to insist upon being heard in the question of university reform. One of the most vigorous protagonists in this matter was Lankester, and his removal to Oxford threatened the little group who had definite notions of academic reform with complete defeat. Luckily Weldon joined us and his energy and enthusiasm were of immense service. We had to fight our own College authorities as well as outside influences. It is not now the fitting occasion to tell the complete history. A joint letter to the *Times* roused the authorities, there were rumours of dismissal from chairs, and of wiser counsels prevailing solely at the instance of a distinguished Liberal statesman then on the College Council. The authorities were supporting a scheme which would have united King's and University Colleges in a second-rate and duplicate London University to be termed the Albert University, and rebellion had to go to extremes, if this project was to be defeated. Weldon with the help of one or two colleagues circularised every member of the College, and the night before the discussion of the charter a widely signed petition against it was in the hands of every member of the House of Commons; the Albert charter was dead, and the College Council hopelessly defeated.

The destructive attitude was now dropt; at a meeting in Wimpole Street, Weldon, G. Carey-Foster and the present writer drafted the scheme, afterwards accepted with small modifications, of an " Association for Promoting a Professorial University for London." The idea was to bring all the London teachers into one camp, to get them to accept a common ideal, and to enlist support for it among thinking men outside. The ideal was the foundation of a university in which teaching should be done by the university professors only, who should largely control the university; the separate colleges were to be absorbed. The aim was thus expressed:

"The creation of a homogeneous academic body with power to *absorb*, not to federate existing institutions of academic rank, seems the real solution of the problem. An academic body of this character might well be organised so far as teaching is concerned on the broad lines of a Scottish University. Such a corporation may be conveniently spoken of as a professorial university to distinguish it from a collegiate or federal university."

The Association met a real need, the London teachers to our surprise and joy joined readily. We got the support of great names in literature and science. We produced a distinct effect on public opinion and by our witnesses even on the Royal Commission. But we considered that we ought to have a leader of great name, and we asked Huxley to be president. Huxley accepted, and came to us with views diverging to some extent from those of the initiators of the Association. Instead of holding up an ideal of academic reform, his plan was to find the minimum which would be accepted by various opposed interests and compromise on the basis of this. The alternatives were a long campaign to impress the powers that be with true notions of academic life, or the immediate acceptance of a teaching university, which should be an *omnium gatherum* of all the teaching institutions in London. The present writer resigned the secretaryship of the Association, and was succeeded by Weldon. It was only after very anxious consideration that the open letter of the former to Professor Huxley of December 3, 1892, was sent to the *Times**. It was a course which Weldon strongly condemned.

---

* *Personal* requests to join the Association had been made to many on the basis of a circular containing the words cited above, the spirit of which was directly repudiated by Huxley.

At a general meeting of the Association held on December 21, 1892, the report of the Executive Committee was received, and after a strong speech from Huxley, adopted. It was then moved by Pearson, and seconded by Unwin : "That the Association trusts that its Executive Committee will persevere in its efforts to establish as far as possibly may be a professorial as distinguished from a federal university." This was carried. At the meeting of the Executive Committee on January 24, 1893, the President presented his own scheme for a teaching University for London ; a vague motion to prepare a scheme to be submitted to the Association was, at the instance of Pearson, seconded by Lankester, amended as follows : "That the Committee prepare a scheme to be submitted to the Association in general accordance with the proposals adopted by the Association." This was carried. On January 25, 1893, Huxley wrote withdrawing his scheme on the ground that the amendment moved by Pearson and seconded by Lankester was "incompatible with any progress towards attainable ends." At the following meeting of the Committee in February, Professors Carey-Foster, Rücker, and Pearson were asked to prepare a scheme embodying the principles of the " proposals" of the Association as a basis for the charter of the proposed university. Mr Dickens and Professor Weldon were added to this committee. The scheme was actually prepared and Weldon aided with yeoman service in the drafting of it*. But the influence of the Association was dead; it never recovered from the divisions thus manifested in its executive. The spirit of compromise and the fatal easiness of federation dominated the situation and the present University emerged out of the chaos. No one felt more bitterly than Weldon the contrast between the original ideal and the result achieved. In fact, it is not too much to say that the greatest hopes for the University, and its most progressive steps since its incorporation, lie in the endeavours made to carry out in part the ideal of a homogeneous professorial university, as it was originally developed one Sunday evening in the house in Wimpole Street, and later substantially reproduced in the proposals of the Association.

This account of one movement, however, with which Weldon was closely concerned would not be justified here, did it not illustrate strongly a marked characteristic of the man. He found his great leader attacked, as he and some others believed, unwarrantably. He wrote one very strong private letter on the point, and never referred to the matter again ; not the slightest breach was caused in his friendship, and the biometric talks, the common work and plans for work were resumed a day or two afterwards as if no source of friction had for a moment arisen. Yet Weldon always felt deeply, and felt this attack on Huxley more than many men would feel a direct personal attack on themselves.

With the death of Huxley in 1895, the Association practically came to an end. Weldon succeeded his hero in 1896 as Crown nominee on the Senate of the University ; here, as on the Board of Zoological Studies of the later reconstituted

* The scheme was printed and adopted by the Association, March 23, 1893.

University, he continued to work and fight for truer ideals of academic administration.

As an administrator and committeeman Weldon combined geniality with strong convictions; he saw at once through flimsy pretexts, and expressed clearly and concisely his own point of view,—"An impulsive loveable man going to the heart of any subject immediately, and always speaking up with great feeling for what he thought right," is how one of his former colleagues aptly describes him. But he lacked several of the essentials which go to make the completely effective committeeman. He was always full of the current piece of research and he grudged all time taken from it; to carry through his own projects he did not adopt the manner of the bull and crush down all opposition; some few men can do this, but it needs not only physique, but its combination with very dominant intellectual power; nor had he the persistency of the corncrake, to wear down his colleagues by continual nagging; nor silent in committee would he molelike be active underneath, "lobbying" his men, and thus more effectually work his will. These types I have known and each was less loved, but more successful than Weldon. He "played the game," threw firmly and well the lance for the cause he thought right, and went his way. He remained to the end the public school or 'varsity lad, whom the idea of "good form" controlled; but unfortunately the type is not so persistent in practical life that it dominates scientific or academic politics. From this standpoint Weldon's death removed from the field a healthy administrator, who acted as a tonic upon weaker colleagues. It was in this sense that he did excellent work, not only on various bodies connected with University College and the University of London, but on the Council of the Royal Society (1896–8), and on its Government Grant and Sectional Committees.

To the biometrician, perhaps, the most interesting committee with which Weldon was associated in these years was that which came later to be called the Royal Society Evolution (Animals and Plants) Committee. It is somewhat difficult to give the full history at present, but some attempt at a sketch of Weldon's connection with it must be made here. Weldon's papers on variation and correlation in shrimps and crabs had brought him closely in touch with Francis Galton, and both were keenly interested in the discovery of further dimorphic forms such as had been suggested by the frontal breadths of the Naples crabs. Weldon was full also of other ideas ripe for investigation. He had started his great attempt at the measurement of a selective death-rate in the crabs of Plymouth Sound; experiments on repeated selection of infusoria were going on in his laboratory; he was gathering an ardent band of workers about him, and much seemed possible with proper assistance and that friendly sympathy which was ever essential to him.

The idea that a group of men can achieve more than a single investigator, if true in some forms of social work, is rarely applicable to scientific committees; but such committees have often been tried in the past, and will no doubt be again attempted in the future. If used as instruments of research, the work done is too often a compromise between different methods and divergent personalities; if merely administrative they are successful or not, according to the width of view of some dominating temperament. If run in the interests of one school, still more of one individual, a committee may no doubt do good work, but it is likely, at the same time, seriously to damage the reputation of any larger body in whose name it works, by too markedly connecting it with one aspect of a problem or one side of an unsettled controversy. These difficulties of the situation seem only by degrees to have come home to the founders of the Evolution Committee.

The project was first discussed informally by R. Meldola, Francis Galton, and Weldon, at a meeting held on the 9th of December, 1893, at the Savile Club. Francis Darwin, A. Macalister, and E. B. Poulton had expressed themselves willing to assist such a project. It was settled that a proposal should be made to the Royal Society for the formation of a committee "For the purpose of conducting Statistical Enquiry into the Variability of Organisms," the members suggested being F. Darwin, F. Galton, A. Macalister, R. Meldola, E. B. Poulton, and W. F. R. Weldon, to whom "it may afterwards be desirable to add a statistician." It was resolved further to ask for a grant of money to obtain material and assistance in measurement and computation.

A Committee* consisting of these members was finally constituted by the Council of the Royal Society, with Francis Galton as chairman and Weldon as secretary, the Committee being entitled: "Committee for conducting Statistical Inquiries into the Measurable Characteristics of Plants and Animals." The use of the words *statistical* and *measurable*, somewhat narrowly, but accurately, defined the proposed researches of the Committee. It went on until 1897, with these members, the same title and scope. Then in the early part of that year its scope was much extended by adding to its objects the "accurate investigation of Variation, Heredity, Selection, and other phenomena relating to Evolution," and W. Bateson, S. H. Burbury, F. D. Godman, W. Heape, E. R. Lankester, M. Masters, Karl Pearson, O. Salvin and Thiselton-Dyer were added to its number. But at present our account must deal with the earlier biometric period of the Committee. Looking back on the matter now, one realises how much Weldon's work was hampered by this Committee. It is generally best that a man's work should be published on his own responsibility, and when he is a man of well-known ability and established reputation, grants in aid can always be procured. In this case Weldon had a sympathetic committee, but the members were naturally anxious on the one hand for the prestige of the Society with whose name they were associated, and secondly, they were desirous of showing that they were achieving something†. Both conditions were incompatible with tentative researches such as biometry then

---

* First meeting, January 25, 1894.

† "Of course these considerations only make the problem more interesting than it was before: and I very much want to solve it. But the committee may say that it requires a problem which is reasonably certain to yield an adequate solution in a fairly short time, and that so risky an attempt as this is not suitable for its present work." Letter of Nov. 13, 1894, relating to the secretion of a specific poison by *Daphnia*.

demanded. Trial and experiment were peculiarly needful in 1893; the statistical calculus itself was not then even partially completed; biometric computations were not reduced to routine methods, and the mere work of collecting, observing, experimenting, and measuring was more than enough for one man. Weldon with his "volume of life" was eager to do all these things, and run a laboratory with perhaps sixty students as well. He was impatient because the probable errors of biometric constants, on which tests of significant differences depend, were not at once forthcoming; he wanted the whole mathematical theory of selection, the due allowances for time and growth, the treatment of selective death-rates and the tests of heterogeneity and dimorphism settled in an afternoon's sitting. The Committee did not possess a mathematician to put on the brake, and Weldon attempted too much in too short a time. Each week Weldon had new and exciting problems, he thrust them upon his friends, demanded solutions, propounded solutions, and was never discouraged when difficulties were pointed out and time asked for.

One of the first subjects to be taken up by the new Committee was to test whether the method of resolution into two Gaussian curves, which suggested dimorphism in the Naples crabs, would be helpful in confirming a similar dimorphism said to exist in the herring. Several thousand herrings arrived at University College, a measurer was trained to deal with them, and the variability of a wide series of characters determined. The distributions came out skew, and Weldon was intensely hopeful that statistical evidence of dimorphism would be forthcoming. Instead of this, the analysis showed dimorphic Gaussian components to be impossible. This result was a great disappointment to him, and, I believe, to the Committee. I could never understand why. A most extensive and valuable series of measurements had been made, which in themselves were well worth publishing. It had been shown that simple dimorphism of a Gaussian kind certainly did not hold for these herrings; in all probability it was a typical case of skew frequency, which would have been most valuable as adding to the known instances, and aiding statisticians eventually to classify such occurrences. But Weldon, and, I presume, the Committee were disheartened, they had been searching for dimorphism and had not found it. The herring data were put on one side by Weldon, and as far as I know have never been published. It is much to be hoped that they may some day be resuscitated from the archives of the Committee (16).

The next point that I personally became aware of in relation to the Evolution Committee was Weldon's attempt to solve the problem of subraces in the case of the ray florets of ox-eyed daisies. I am unaware who brought the material before the Committee, but it was obviously heterogeneous in the highest degree. There was no evidence at all that any attempt had been made to allow for seasonal and environmental effects, and whatever truth there may be in a tendency of the modes to fall into Fibonacci groups, we now know that varying season and period will produce within a certain range almost any mode in this flower*. To break

* *Biometrika*, Vol. I. pp. 305, 309 *et seq.*

up such a heterogeneity even into Gaussian components was a problem not then solved, and one which has not since been solved. It was cruel fate that thrust such a problem on Weldon, and kept him over it for weeks. He was struggling with most highly complex mathematical difficulties, and actually beginning with a problem which a more highly trained mathematician would certainly have put on one side in the then state of statistical analysis*.

The next portion of the Committee's work was far more successful—the "Attempt to Measure the Death-rate due to the Selective Destruction of *Carcinus moenas*, with respect to a Particular Dimension" (17). This formed the first report of the Committee, and was presented to the Royal Society in November, 1894. Weldon's general project in this case was, I believe, absolutely novel at the time, and embraces, I consider, the best manner still of testing the truth of the Darwinian theory. It consists in determining whether the death-rate is correlated with measurable characters of the organism, or, as he himself put it, " in comparing the frequency of abnormalities in young individuals at various stages of growth with the frequency of the same abnormalities in adult life, so as to determine whether any evidence of selective destruction during growth could be discovered or not."

Thus stated the problem might appear an easy one, but it is the very reverse. How is the 'abnormality,' i.e. what we should now term the deviation from type, to be measured at each stage of growth ? What is to determine 'adult' life ? What measure is there of the time during which the individual adult life has been exposed to the selective destruction ? Weldon undoubtedly chose the crab because of the facilities it offers for measurement. But its age then becomes an appreciation based merely on the obviously close, but probably imperfect correlation between age and size. Further, the law of growth, complicated rather than simplified by the moults, and the question as to how far the variability of the characters dealt with is affected by growth combine, in the case of crabs, to form an exceedingly difficult problem. It is practically impossible to keep a sufficiently large series of crabs through the whole period of adolescence, and if it were possible, it is far from certain that the claustral environment necessary would not sensibly affect their law of growth.

Looking back now on Weldon's paper of 1894, one realises its great merits ; it formulates the whole range of problems which must be dealt with biometrically before the principle of selection can be raised from hypothesis to law. Almost each step of it suggests a mathematical problem of vital importance in evolution, which has since been developed at length, or still awaits the labour of the ardent biometrician. On the whole, I think, Weldon came very near to demonstrating his point, but whether he did or did not scarcely affects the suggestiveness of the paper†.

---

* We now know that some of the most skew distributions are given by the parts of flowers, and the problem propounded to Weldon was to resolve into a number, probably five or six, of such skew components a strikingly irregular frequency distribution for ray florets !

† Reading through the criticisms I communicated to him at the time, criticisms written purely from

Unfortunately the paper, as well as the suggestive " Remarks on Variation in Animals and Plants " (18), with its memorable words :—" The questions raised by the Darwinian hypothesis are purely statistical, and the statistical method is the only one at present obvious by which that hypothesis can be experimentally checked "—fell on very barren soil. The paper produced a mass of criticism—folios were written to the Chairman of the Committee, showing how this, that or the other vitiated entirely the results. The very notion that the Darwinian theory might after all be capable of statistical demonstration seemed to excite all sorts and conditions of men to hostility. Weldon, instead of being allowed to do his own work in his own way, had to be constantly replying to letters, some even eighteen sheets long, addressed to the Chairman of the Committee. These letters were not sympathetic and suggestive, but mostly purely controversial. The need for further investigation of the law of growth had been frankly admitted by Weldon in the " Remarks " issued at the discussion on the " Report," but the critics declined to wait for answers till further results were published. This attack lasted for the next three years, during which further researches on the selective death-rate and growth of crabs were carried out, and it formed a serious impediment to calm progressive investigation. A further instructive report (19) on the growth at two moults of a considerable number of crabs was made to the Committee in 1897, but I believe has never been published. Later, an account of work on Natural Selection in crabs was given by Weldon in his " Presidential Address to the Zoological Section of the British Association," Bristol, 1898 (20).

In this paper Weldon returns to the problem of whether frontal breadth in crabs is correlated with a selective death-rate, but he now deals with type and not variability. He first approaches the problem from the consideration of whether for this character the crabs in Plymouth Sound are remaining stable, and he shows from measurements made by Sir Herbert Thompson and himself during the years 1892 to 1896, that the population is unstable. He next seeks a cause for this secular change, and he finds it in the turbid state of the water in Plymouth Sound, due to the continual carriage into it of large amounts of china clay and sewage. Direct experiments were then made on the selective death-rate of crabs kept in water with suspended china clay and on another occasion in foul water. In all cases the survivors were found to have a smaller frontal breadth relatively to their carapace length. Confirmatory experiments showed that after the first shock of confinement was passed this selection did not occur among crabs kept in pure sea water. A reasonable explanation of this selective action was provided in the character of a crab's breathing apparatus. Thus, after several

the mathematical standpoint, I still think them valid, but I realise also how much of my own work flowed directly from the suggestiveness of this paper. In fact it was the starting-point of the whole of the work on the influence of selection on the correlation and variability of organs. The sequel to that work, the influence of selection during growth, flows equally from Weldon's paper, but although we know much more than we did ten years ago as to the laws of growth, no sufficiently general formula of growth can yet be applied to allow of the completion of Weldon's work in this direction.

A "Crabbery" at Plymouth.

years of discouragement and much hard labour, Weldon succeeded in demonstrating that natural selection was really at work, and further that it was at work at a very sensible rate*. The labour involved was excessive. One "crabbery" consisted of 500 wide-mouthed bottles, each with two syphons for a constant flow of sea water, and each crab had to be fed daily and its bottle cleaned. During the summer of 1897 Weldon spent the whole of his days at the aquarium, and his wife hardly left him except to fetch the needful chop. The sewage experiment was "horrible from the great quantity of decaying matter necessary to kill a healthy crab." In 1898 the china clay experiments were continued at Plymouth. But in the autumn a rest came. The Address was written and Weldon thoroughly enjoyed his presidency of Section D of the British Association at Bristol.

It may not be out of place here to note the great aid Weldon's artistic instinct and literary training gave to his scientific expression. His papers are models of clear exposition, his facts are well marshalled, his phraseology is apt, his arguments are concise, and his conclusions tersely and definitely expressed. The result, however, was not reached without much labour. I do not mean that it was an effort to him to write well and clearly, but that his standard was so high, that having written a memoir, he would to please his own sense of the fitting rewrite the whole of it and possibly redraw all the diagrams. Nor was the remodelled memoir necessarily in its final form. A third or fourth reconstruction might follow to satisfy his own standard of right expression. To him a paper was a literary whole, which had not only to convey new facts, but to play its part on the scientific stage,—and he was not satisfied until it was in his judgment artistically complete. There was never any artificial brilliancy introduced in the process; rhetoric in the service of science was intolerable to Weldon. It was simply an attempt to choose the suitable form and the right words for a given purpose. It was comparable with Weldon's sense of sound, with his extraordinary gift of appreciating and reproducing the exact intonation of a foreign tongue. Both were the result of observation and experiment—not manifest in the final product—guided by a trained artistic sense.

Considerable changes were soon to take place in Weldon's environment and scheme of work. Lankester had been appointed director to the British Museum (Natural History), and in February, 1899, Weldon succeeded him in the Linacre Professorship at Oxford. In the February of 1897 the Royal Society Evolution Committee received a large increase of membership; it ceased henceforth to "conduct statistical inquiries into the measurable characteristics of plants and animals." It became transformed into an Evolution (Plants and Animals) Committee. At first there were great hopes of achievement, there was a possibility of securing Charles Darwin's house as a centre for breeding experiments, and a considerable sum of money was promised in aid. Francis Galton struggled bravely for a great idea. He wanted to see the numerous bodies engaged in horticulture

---

* The 60,000,000 years or so, which the physicist then allowed the evolutionist, were at that time a little more of an incubus than they are now!

and zoology coordinated in at least one aspect of their work, and that research of a scientific kind should be introduced into the proceedings of each of them. He strove to make two schools, widely diverse in method and aim, understand each other. He wanted to keep individuals and societies up to their work, and prevent overlappings. But it was not to be. The members were pulling in opposite directions, there was too much friction, and too little compromise. A false antithesis was raised between what was termed "natural history" and any sort of statistical inquiry leading to numerical results. The biometric members ceased to attend regularly and finally resigned towards the end of 1899. Thiselton-Dyer and Meldola also left the Committee, which became from that date confined to one special school and one limited form of investigation. From beginning to end the Committee has, in the opinion of the present writer, been a mistake; not only because at first it distinctly forced the pace and hampered Weldon's work, but because experience shows that such a committee can only work effectually in the interests of one school of ideas, and this, whatever safeguards may be taken, has at least the appearance of destroying the impartiality of the parent body, a matter of very grave importance.

During the eight years of Weldon's London professoriate his development was great; he became step by step a sound mathematician, and gained largely in his power of clear and luminous exposition. His laboratory was always full of enthusiastic workers, and over forty memoirs were published by his students, who included E. J. Allen, E. T. Browne, F. Buchanan, G. H. Fowler, E. S. Goodrich, H. Thompson, E. Warren, and others of known name. The following lines, provided by a friend, graphically recall Weldon in his early London days :—

" In so vivid a personality it is hard to point to the period of greatest mental activity, but of the nineteen years in which I knew him I should select the first few years of his Professorship at University College, London. Fresh from contemplative research at the Plymouth Laboratory of the Marine Biological Association, and with his mind full of the new problems to which the study of marine life had introduced him, he threw himself into teaching with renewed zest. The effect on his students was amazing ; most of them began a zoological course as a compulsory but annoying preliminary to a degree ; Weldon soon changed that. Without ever forgetting the requirements of examinations, he made the subject alive and absorbing ; his advanced classes soon filled up ; and while on the one hand, the scholarships at London University were always claimed by his students, on the other the output of original investigation published by his department was one of which no university need have felt ashamed. Besides all this his students loved him ; he was so intensely human....Into the question of remodelling the University and the defence of his College, Weldon threw himself as if unencumbered with arduous teaching and research ; his notably lofty ideals and vigorous championship were far from being wasted ; but his removal to Oxford at the time of the birth of the new University was a severe loss to the cause of real education in London. Gentle with ignorance, he was fiercely intolerant of educational shams and cants."

As the present writer has indicated, the stress during these London years was very great—the struggle with new mathematical processes, the wear of incessant calculation, the worry of unending controversy to a man fully occupied with research and teaching, all told on Weldon. The holidays were more limited in extent, but were very varied in character. In 1893, Easter was spent in the Sieben Gebirge; the Weldons were up at six, calculating till one, and starting a great tramp at two, from which they returned at eight. The autumn they spent in Venice, going by sea, and the Christmas at Brussels, with opera each night and walks to Waterloo most days. In 1894 it was Siena for the Palio, with a knapsack tour from Stresa to Alagna by Orta and Varallo. In 1895 Easter found them fossil-collecting in the Eiffel, and, after the hard summer at Plymouth, in the Apennines, winding up in Florence. Bicycling was the rule in 1896, even cycling from Wimpole Street to Plymouth, and the only holiday a cycling tour in Normandy. In 1897 there was an Easter visit with architectural sketching to the cathedrals of North-East France, and after the specially hard summer at Plymouth a trip to Perugia and a return from Genoa by sea. The last year in London included a butterfly and moth collecting expedition to Ravenna at Easter, but no summer holiday abroad; the British Association, followed by a study of Wells Cathedral, occupied its place. The restlessness of work seemed to have overflowed into the holidays, and Weldon's friends knew that it was telling upon him, and trusted that Oxford life might be quieter than the London life had been.

VI.   *Oxford and the Second Professoriate, 1900—1906.*

The removal of Weldon from the London field of work, while an incalculable loss to his colleagues, was not without compensation to his nearest friends. They knew that the life of the last few years had been one of great tension, that Weldon's time had been too much encroached upon by committee work, that the separation between the locus of his teaching and of his research work was very undesirable, that even the social life of London involved too much expense of energy. Oxford, in some respects, would present a narrower field of administrative duties; it would provide a roomy and amply equipped laboratory, where experiments hitherto shared between Plymouth and Gower Street could be carried out, and remain under control while ordinary teaching work was going forward. Even the social life in Oxford had more regular hours and was less over-stimulating. It is true that Weldon occasionally regretted the contact with many minds working on kindred topics, and even the stimulus of keen men working on quite different subjects, which is characteristic of the metropolis. He would speak with great affection of "dear old Gower Street, where everybody was working and everybody wanted to work"—and he would be vexed that so many of "these nice Oxford boys" had no *res angusta domi* to force them from the river and the playing field into the laboratory and the lecture room. "They are so nice, they come to my lectures because they think it would be rude to leave me alone." The lad who would not make a sacrifice to his love of science—accept an Asiatic appointment of the merest bread and butter value, or take passage in a tramp steamer to collect in South America—was anathema to him. He wanted everywhere an infant Huxley, realising the value of tropical or semi-tropical observation and experience and anxious to seize the opportunity of it at any slight personal inconvenience. Weldon did not grasp that it was largely his own personality which had created the band of earnest workers round him in London, and that with time it would be effective in more conservative Oxford. He did not realise that the over-stimulus of the London period, with its midnight hours and incessant interchange of ideas, would be better replaced by the more leisurely intellectual and more regulated social life of Oxford.

There is another point which emphasises the value of this change. Weldon's taste, his whole emotional nature, made him essentially a field naturalist. It was no innate taste for figures or symbols, no pleasure in arm-chair work, which drew him to statistical research. Nor was it the influence of any personality. On the contrary, he was impelled to it by the feeling that no further progress with Darwinism could be made until demonstration from the statistical side was forthcoming. His biometric friendships arose from the direction he felt his work must take. He distrusted mathematicians as much as any good Mendelian might do; they were persons who neither observed nor experimented, who had "a true horror of a real measurement." Acceptance of each stage of biometric theory could only be won from Weldon by a tough battle; it had first to justify its necessity, and next to justify its mathematical correctness. He was not drawn into actuarial work by his sympathies or his friendships, he was *driven* into it by the looseness he discerned in much biological reasoning; he felt an *impasse*, which could only be surmounted by the stringency of mathematical logic. Those who have known Weldon collecting on the shore, dredging at sea, or in later days sampling ponds and wells for his *Crustacea* book, photographing snail environments in Sicily, or hunting for *Clausilia* in the woods at Risborough or Plön, realise that he was in the first place the open-air naturalist. If further evidence be indeed wanting, let the following words provide it:

[*April*, 1903.]

"Just back, and have just read your letter. I will play with the spanner and talk of it to-morrow.

I did not telegraph because our office was shut. It was a great disappointment to miss you; but the ride was the one thing I enjoyed out of the last three weeks. I have felt nothing like it since the old days when I used to lie in a fishing boat dodging the squalls off Rame Head or the Deadman, when we were all young and arithmetic was not yet. That is all gone. The good old man I used to sail with went to haul lobster pots in one of the March gales, and his boat was found bottom upwards.

He was a good soul. 'Yes, my dear,' he used to say in a breeze; 'we'll shake out all them reefs if you like. You'll get wet, but I'm only a fisherman and wet don't hurt me.' Then he would sing Devonshire songs while the water came over the gunwale, till you went on your knees to him to ask for at least one reef back again.

Really, even Basingstoke railway station looked good with the squalls climbing round it. *Ride* home. It will do you no end of good. Go by Farnham, Basingstoke, not by Guildford. Sandro and I rode home to-day. We had no snow, and no rain, and not half the fun of Monday. A sober, middle-aged ride on a good road in good weather.

Nevertheless, my head is so full of chalk-downs and clouds, and things, I can't write biometry to-night. Always, when I have been with the country, the feeling breaks out that the other folk have the best of it. The other way you live with the country and become part of it; and you dredge, or fish, or shoot something wonderful, and you describe it, and everyone sees that it is wonderful, and you all enjoy the wonder. And there is no solution, and if there were, it would not be worth the shadow of a shower flying across the country.

And this is all wicked nonsense, and I am going to bed. Yours affectionately,

W. F. R. WELDON."

Weldon was a child of the open air and the breezes, and we hoped that he might have more of them, if not in lowland Oxford, at least on the hills around. There was space and air too for the experimental work that had been so cramped in Gower Street. The *Daphnia* studies, which had occupied so much energy under unfavourable conditions in London, were at once resumed on broader lines in the ponds and ditches round Oxford. Weldon, with a basket of bottles attached to his cycle handle, and a fishing creel, filled with more bottles, on his back, might be met even as far as the Chilterns, collecting not only *Daphnia*, but samples of the water in which they lived. His University College work had shown him how widely *Daphnia* are modified by their chemical and physical environment, and how this modification is largely due to selection. There exist elaborate drawings of the *Daphnia* from the Oxfordshire ponds, indicating their differentiation into local races, and notes on the peculiarity of their habitat and the chemical constitution of the water:

"In the meantime I have been led into a non-statistical work for the moment. Get out of the library and read Klebs : *Bedingungen der Fortpflanzung bei einigen Algen und Pilzen.*

By tricks of nutrition, light, etc., Klebs can make simple algae reproduce either a-sexually or sexually, or parthenogenetically, as he pleases. In cases where every textbook tells you that a regular alternation of sexual and a-sexual generation is the rule, he can make *either* form recur as often as he likes.

If one can by similar tricks throw *Daphnia* into this condition, then the measuring machine can again come into play, and one can compare parthenogenetic inheritance with sexual inheritance as often as one pleases.

That is a *Nebensache.*—The *Hauptsache* here is the great variation in the chemical composition (pardon the phrase) of the water in the little rivers. Their percentage of dissolved salts varies enormously, and I hope to go about as I have begun, with a large fisherman's creel tied to the handle-bar of my bicycle, learning the correlation between the salts in the waters and the fauna.—Then again comes the measurement, and the attempt to derive one local form from the other under controlled conditions by direct selective destruction due to the conditions."

This was precisely the same problem which a study of Kobelt's *Studien zur Zoogeographie*, 1897–9, led him later to take up with regard to land snails. What is the meaning of the slight but perfectly sensible differences in type to be found in shells from adjacent valleys, or even from different heights of the same mountain ? Weldon attacked the problem in his usual manner; he spent two Christmas vacations collecting Sicilian snails of the same species from habitats extending over a wide area, the local environments were described, and the snails often photographed with their immediate surroundings. Innumerable shells were brought back to Oxford, and Weldon delighted to discourse on the significant differences in local type, and yet the gradual change of type to type from one spot to another.   No rapidly made measurement on the outside of the shell would satisfy Weldon ; the shell must be carefully ground down through the axis, and measurements made on the section thus exposed.   Perhaps four or five snail shells could be ground and measured in a day, and at the time of his death, not more than a few hundred of the Sicilian thousands had even been ground.   Like the *Daphnia,* the Sicilian snails remain as an indication of the way—the path of absolute thoroughness—the master would have us follow.   " Life is not long enough for biometry," murmurs the superficial critic.   But the man of deeper insight replies :

> That low man goes on adding one to one,
> His hundred's soon hit :
> This high man, aiming at a million,
> Misses an unit.

But these attempts to get to the kernel of selection in its action on local races were far from occupying the whole of Weldon's thoughts in these early days.   In conjunction with·his assistant, Dr E. Warren, he had commenced at University College his first big experimental investigation into heredity.

"The Oxford rivers have had to rest during the last few weeks, because of the pedigree moths. These are apparently going on very well indeed.   There are at present about 3500 caterpillars, belonging to thirty-eight families forming the third domesticated generation."

The characters to be dealt with consisted of the number of scales in particular colour patches, and the work of counting these was very laborious.   A little later (16 July, 1899) Weldon writes :

"The caterpillars are hatching by hundreds and I hope the clean air will help them to do better here than in London.   From egg to moth, poor Warren, in spite of magnificent efforts, had a death-rate of over eighty per cent. ; and that seems to me a rather serious thing…because one cannot be sure that the death-rate was not partly selective with respect to things in the caterpillar which are correlated with colour in the moth.   The influence of climate is shown by the fertility of the eggs—Warren got forty per cent. of fertilised eggs from his pairing and an average of over one hundred eggs per batch.   Nearly all my pairs lay fertile eggs and those I have counted give an average of one hundred and sixty-five eggs per batch."

And again, on the 14th August of the same year :

"I want to come and talk to you, especially about death ; but I cannot come till my caterpillars are safely turned into pupae.   For the sake of these caterpillars I have, at the risk of personal liberty and reputation, stolen from the roadside one hundred square feet of clover turf, the property of the Lords of various Manors in this neighbourhood.   The little ruffians have now eaten all this clover and for the last day or two of their existence have to be fed by hand.—Therefore I have to pluck fresh clover (which is·not stealing if you do not do it in an enclosed pasture) every day.—My bicycle is nearly worn out from carrying extra weight.   Riding down a steep hill, with your brake smashed, and with five or six feet of heavy turf on your back is like playing at Attwood's machine.   You get very near to the theoretical acceleration too ! "

In the course of three years many hundreds of pedigree moths were dealt with and the observations were reduced.   But *no definite inheritance at all of the character selected for consideration was discovered.*   Weldon, I believe, thought that there had been some fatal mistake in the selection of pairings, and undoubtedly in some cases parents of opposite deviation had been mated, so that a rather influential negative assortative mating resulted. But from other series of pedigree moth data that I have since seen, it seems to me probable that there is some special feature in heredity in moths, or possibly in those that breed *twice* in the year, and that the vast piece of work which Weldon and Warren undertook in 1898—1901, may still have its lesson to teach us. At the time it formed another link in that chain of apparent failures which for a time, but only for a time, disheartened Weldon.

In these three first years at Oxford, Weldon's intellectual activity was intense. The letters to the present writer, which in 1899 averaged one a week, in 1900 and 1901 reached an average of two, and in some weeks there were almost daily letters. These letters not only teem with fruitful criticism and suggestion with regard to the recipient's own work, but contain veritable treatises—drawings, tables, calculations—on the writer's own experiments and observations. To the pedigree moth experiments was added in the summer of 1900 an elaborate series of Shirley Poppy growings, 1250 pedigree individuals being grown and tended in separate pots; Weldon's records were the most perfect of those of any of the cooperators, and his energy and suggestions gave a new impetus to the whole investigation. They were ultimately published in *Biometrika* under the title, *Cooperative Investigations on Plants,* I. *On Inheritance in the Shirley Poppy* (22). As Weldon himself expressed it, the moths and poppies meant "a solid eight hours daily of stable-boy work through the whole summer, and through the Easter vacation, with decent statistical work between." The autumn of 1899 provided no proper holiday, but Christmas found the Weldons in Rome. After the Shirley Poppies were out of hand in the summer of 1900, the Weldons went to Hamburg and thence to Plön. The object of this visit was to collect *Clausilia* at Plön and Gremsmühlen for comparison with the race at Risborough. The same aim—the comparison of local races—led Weldon at Christmas to collect land snails in Madeira. Thus he slowly built up a magnificent biometric collection of snail shells—i.e. one sufficiently large to show in the case of many local races of a number of species the type and variability by statistically ample samples. Of this part of Weldon's work only two fragments have been published, "A First Study of Natural Selection in *Clausilia laminata* (Montagu)" (23), and "Note on a Race of *Clausilia itala* (von Martens)" (24). In the first of these memoirs Weldon shows that two races of *C. laminata* exist, in localities so widely separated as Gremsmühlen and Risborough, with sensibly identical spirals, although no crossing between their ancestors can have existed for an immense period of time, and although there are comparatively few common environmental conditions. At the same time, while no differential secular selection of the spiral appears to have taken place during this period, there yet seems to be a periodic selection of the younger individuals in each generation, the variability of the spirals of the young shells being sensibly greater than that of the corresponding whorls of adults. In other words stability to the type is preserved by selection in each new generation.

In the second memoir Weldon sought for demonstration of a like periodic selection in the *C. itala* he had collected from the public walks round the Citadel of Brescia. He failed, however, to trace it, and was forced to conclude that *C. itala* is either not now subject to selective elimination for this character, or is multiplying at present under specially favourable conditions at Brescia, or again, as both young and old were gathered in early spring, after their winter sleep, that elimination takes place largely during the winter, and "that individuals of the same length, collected in the autumn, at the close of their period of growth, might be more variable than those which survive the winter."

Quite apart from the results reached, Weldon's papers are of the highest suggestiveness. Does selection take place between birth and the adult or repro- ductive stage? This is the problem which everyone interested in Darwinism desires to see answered. But to answer it we need to compare the characters of the organisms at the same stage of growth, for these characters are modified by growth. How is it possible to compare a sample of the race at an early stage, with its adult sample? The problem of growth, to be studied only under conditions of captivity, possibly modifying the natural growth immensely, had made the crab investigation an extremely complex one. Weldon solved the difficulty by the brilliant idea that the snail carries with it practically a record of its youth. If the wear and tear of the outside of the shell to some extent confuses the record there, a carefully ground axial section will reveal by the lower whorls the infancy of the organism. Hence the days given to experimental grinding, the training in manipulation and the final success, and then the steady work, grinding and measuring a few specimens a day, till the necessary hundreds were put together; the laborious calculations not in the least indicated in the papers—the arithmetical slips with bad days of depression, and the completed result: the illustration of how shells may be used—by those who will give the needful toil—to test the truth of the Darwinian theory.

The summer of 1899 found the present writer at Great Hampden on the Chilterns, working at poppies and developing a theory of homotyposis, namely, the quantitative degree of resemblance to be found on the average between the like parts of organisms. Weldon, who came over from Oxford to dredge the ponds and to discover *Clausilia* by the White Cross at Monks Risborough, provided the criticism, suggestion, and encouragement, in which he never failed* :

"You have got hold," he wrote, after returning to Oxford, "of the big problem which all poor biologers have been trying for ever so long. I wish you good luck with it."

The collection and reduction of material were on a larger scale than had been previously attempted, and the memoir was not presented to the Royal Society until the October of the following year (1900). It was soon evident that the attitude of the Society with regard to biometry was undergoing considerable change. The meeting of November 15 and the discussion that took place on the

---

* His aid in the second part—Homotyposis in the Animal Kingdom, shortly to be published,—was even more substantial.

homotyposis paper was the immediate cause of the proposal to found this Journal. A little later a detailed criticism of the paper by one of the referees was actually printed by the Secretaries and issued to Fellows at a meeting, before the fate of the criticised paper itself had been notified, and before the paper itself was in the hands of those present. This confirmed the biometric school in their determination to start and run a journal of their own.

On November 16 Weldon wrote:

"The contention 'that numbers mean nothing and do not exist in Nature' is a very serious thing, which will have to be fought. Most other people have got beyond it, but most biologists have not.

Do you think it would be too hopelessly expensive to start a journal of some kind?......

If one printed five hundred copies of a royal 8vo. once a quarter, sternly repressing anything by way of illustration except process drawings and curves, what would the annual loss be, taking any practical price per number?... If no English publisher would undertake it at a cheap rate, the cost of going to Fischer of Jena, or even Engelmann, would not be very great."

This was the first definite suggestion of the establishment of *Biometrika*. On November 29, the draft circular, corresponding fairly closely to the first editorial of the first number (25), reached me from Oxford with the words: "Get a better title for this would-be journal than I can think of!" The circular went back to Oxford with the suggestion that the science in future should be called Biometry and its official organ be *Biometrika*. And on December 2nd, 1900, Weldon wrote:

"I did not see your letter yesterday until it was too late for you to have an answer last night. I like *Biometrika* and the subtitle. Certainly we ought to state that articles will be printed in German, French, or Italian. One may hope for stuff from anthropologers, and —— for instance, ought to be allowed to use his own tongue."

Thus was this Journal born and christened. The reply to circulars issued during December was sufficiently favourable to warrant our proceeding further. A guarantee fund sufficing to carry on the Journal for a number of years was raised at once; good friends of Biometry coming forward to aid the editors. By June of 1901 its publication through the Cambridge University Press had been arranged for, and the sympathetic help of the Syndics and the care given by the University Printers enabled us to start well and surmount many difficulties peculiar to a new branch of science*. Those of us who believe that *Biometrika* came to stay and to fulfil a definite function in the world of science hope that the name of the man who first formulated a definite proposal for a biometric organ may always continue to be associated with our title-page. During the years in which Weldon was editor he contributed much, directly and indirectly, to its pages. He was referee for all essentially biological papers; and his judgment in this matter was of the utmost value. He revised and almost rewrote special articles. He was ever ready with encouragement and aid when real difficulties arose. For the mechanical labour

* A special feature has been of course the masses of tabulated numbers. It deserves to be put on record that on more than one occasion 15 or 20 pages of figures have been set up without a single printer's error.

of editing, for proof-reading, for preparation of manuscripts and drawings for the press, or for interviews with engravers, he had little taste or time. He was too full of his current problem to undertake work of this kind regularly; proofs might remain for weeks unopened, until the number was printed off, and manuscript might disappear into a drawer, when the co-editor imagined it was safely on its way back to the author! But Weldon was always delightful when such "laches" were discovered; to meet Weldon when he was in an apologetic frame of mind meant that you must apologise to him yourself for the very thought of scolding him! It was all over before he had shaken hands, sat down, and lighted the inevitable cigarette; you were not talking of proofs, but of Kobelt, Mendel, Maeterlinck, the *Kritik der reinen Vernunft*,—anything and everything but dull editorial matters. And you felt a freshness and a tonic, and a sense of the healthy joy and pleasure of life, and you wondered how it was possible to do anything but love this man and rejoice in the clearness of his vision and the suggestiveness of his thought.

Starting on October 16, 1900, and extending throughout the early *Biometrika* letters, runs a flood of information with regard to Mendel and his hypothesis.

"About pleasanter things, I have heard of and read a paper by one, Mendel, on the results of crossing peas, which I think you would like to read. It is in the *Abhandlungen des natur-forschenden Vereines in Brünn* for 1865. I have the R.S. copy here, but I will send it to you if you want it."   [*October* 16, 1900.]

Then follows a resumé of the first of Mendel's memoirs, and for months the letters—always treatises—are equally devoted to snails, Mendelism and the basal things of life. It is almost impossible to give an idea by sampling of the crush of keen and vital interest these letters represent. Some attempt must, however, be made:

"Have you ever been up here? It is not at all a bad little country when you are tired.—We started simply to see the architecture at Lübeck, because neither of us knew the North German brick and wood church work. That was very interesting, then we came here for fresh air and quiet—and we found SNAILS.

I have rather more than 5000 snails all properly pickled, with localities recorded.......There are so many points about snails, if one could only measure and breed them !...Also I have been greatly impressed with the way in which they are dependent upon conditions of environment, so that one quickly learned to know almost exactly what species one would find in any place one passed through. I think that by going from here, which is almost the eastern limit of several species, to a very different country, such as Oxford, one might almost hope to make a good shot at some of the essential conditions which determine the distribution of some of the species.... It is ridiculous that such abundant material as snails afford everywhere (except at Danby End?) should be left useless because one cannot see how to take advantage of it. Send me some "tips" for trials. (To Oxford,—we go home to-day)."   [Plön, 5/9/90.]

"You ought to see Lübeck some day. You know so much about German art that I suppose the pathetic ugliness of it does not hurt you any more?......

You can't get a beautiful art in a climate where people must wear clothes. Just as the northern idea of a portrait is a round face stuck on top of a heap of fine clothes, so the northern idea of a building is a thing with all its good simple lines disguised by silly excrescences. If you want to see really majestic brickwork go to Siena or Pistoja, where you can see naked men and women in the streets on a summer's day. Lübeck is very earnest, and very interesting, and so on and so on, but it is *not* beautiful !... I am sending you a parcel of snails that you may see the sort of thing.

In one beech wood, on the trunks of the trees, we collected rather more than 3000 snails, most of them *Clausilia biplicata* and *C. laminata* (see the parcel), but some *Helix lapidicida.* There are certainly one thousand each of the two *Clausilia* species from this locality, and four or five hundred of each from another wood."

Then follow pages of minute description of each type of shell in the parcel and discussion as to the possibility, by grinding or by taking a melted paraffin cast of the inside, of measuring biometrically this or that character.

"I fancy want of moisture must have more to do with the absence of snails about you than want of chalk. What are you on ? Surely you have nearly the same Oolites and Lias that we have here ?...... Have you committed the sin of digging up a bit of your moor, and looking among roots ?"   [Oxford, 13/9/1900.]

"A happy New Year to you ! I send in another envelope specimens of the problematic snail, which has been found in sufficient numbers already. The pattern cannot, I think, be treated as due to lines of growth, and I hope it will be possible to find some way of estimating its variability.—It occurs more or less in a whole series of species here, and here only ; and the hills here are so separated by deep valleys, with great climatic differences at different elevations, that there are well isolated local races.

It is rather hard for me to collect many races, because I have to look after my sick laboratory boy, and to teach him sea work, which takes time and produces only isolated examples of pretty museum things, which are a joy to see, although they teach one very little*. To do this and also to find time for a 2000 or 3000 feet climb after snails makes a very good day, and one goes to bed very fit, and full of beautiful remembrances. As one walks up hill, the impression is very absurd. Here the garden is full of bananas and sugar canes ; in an hour's climb one gets into a wood of pine trees and heather, and looks down on to all this tropical valley. The contrast is very curious, and I have not got accustomed to it.

It seems rather a bad year for land beasts. The normal rainfall in December and January is said to be about thirty inches ; and this year practically no rain has fallen since the spring. I suppose as a result of this every live thing gets under the biggest block of basalt it can find. This makes snail hunting rather exciting, because when you get to the top of a kopje where the beasts are you find the sides so steep that any stone you disturb rolls down, unless you take great care. My first day's hunt resulted in such a roll. A stone which I could only just lift rolled down into a sugar cane bed some three or four hundred feet below. I have never felt so ill as I felt until I found that stone and made sure that it had not smashed up an innocent Portuguese peasant !

---

* One of the blows to Weldon, which resulted from his biometric view of life, was that his biological friends could not appreciate his new enthusiasms. They could not understand how the Museum "specimen" was in the future to be replaced by the "sample" of 500 to 1000 individuals, not to be looked at through a glass, but to be handled, used, and if necessary *used up.* They warned his pupils solemnly to give up this sort of fooling and take to the real business of the "biologer," if they wished for success. "I told —— about these snails," Weldon wrote on Oct. 11, 1900, "and he wrote me an earnest letter, urging me to return to the pleasant way of describing beasts for the delight of the faithful. That is the real thing if you want to be popular. Go to sea, and have a good time, and bring back a jelly fish which is bright blue."

There is much missionary work yet to be done by biometricians, and Weldon's loss will make it still harder !

If one knew anything about natural history one might do a great service to these people. The whole place is covered with the tracks of a little black ant, introduced with some South American sugar canes five or six years ago. The ant cultivates a number of aphides, which produce serious diseases on all the fruit trees ; also it attacks the newly hatched birds and all beasts which shelter under stones. Under a big stone, where some dozen snails have sheltered, about half the shells (which look quite fresh) are eaten empty by ants ; so it is with the beetles, grasshoppers, and other things. The only good thing they have done is to eat the cockroaches. Every kitchen is now full of ants, and contains no cockroaches at all.

How snails make their shells here is hard to understand ; there is not a scrap of limestone in the place : all basalt and beds of gorgeously coloured volcanic sands. Yet when one finds the right place, one finds that snails swarm and their shells are rather harder than usual !

I should very much like to know whether the habit of hiding under stones is as general in all seasons as it is now. You know Wallace points out that most of the beetles here have lost their wings ; and he regards it as probable that flying beetles would be blown out to sea in storms. Now first of all there are practically no storms, and secondly, if there were, the valleys are so deep and their sides so precipitous, that there is abundant shelter against winds. But the loss of wings might well be correlated with the habit of walking under stones to get out of the sun. You find a patch of bare hot sand, so steep that you can hardly stand on it, with a stone here and there, and no sign of any living thing. If you turn over a stone you find a number of snails, a lizard, twenty or thirty beetles, a grasshopper or two, and armies of millipedes.......
The man we see most of in this inn is a splendid creature. A captain in the Canadian frontier police, who volunteered for service in South Africa, and is recovering from a bullet through his right lung. Because he has a colonial accent, —— cannot see any merit in him....... It is only another sample of the difficulty I feel every day at Oxford. The boys there are so occupied with silly superficial things that one can never bring them to think of fundamental matters."

[Funchal, 29/12/00.]

"I am glad you are disgusted with the Life. I was afraid you were not.—You cannot judge the man from the bits of his letters. I do not think one ought to try to have an opinion about a man's conduct towards his wife, or indeed about his ethical value at all. One cannot possibly get hold of evidence enough, and the little bits of bad journalism which people give one are only sufficient to disturb one's mind. Take the old man as one knows him by his work, without troubling to guess at his motives, and there is not much the matter with him. I quite agree with you in loving Darwin and —— more ; but a man may be a great deal lower than these two, and yet be high enough for reverence." [Oxford, 2/12/00.]

The earlier part of 1901 was chiefly occupied by snails, but a new factor had come into Weldon's many-sided occupations. It was settled that *Biometrika* should have in an early number a critical bibliography of papers dealing with statistical biology. Weldon undertook this task as his study of Mendel had led him to a very great number of such papers dealing with inheritance, and the section on Heredity was to be published first. Like all Weldon's projects, it was to be done in so thorough and comprehensive a manner that years were required for its completion. A very full list of titles was formed, especially in the Inheritance section, and many of the papers therein were thoroughly studied and abstracted (26). But such study meant with Weldon not only grasping the writer's conclusions, but testing his arithmetic and weighing his logic. Thus Weldon's Note on "Change in Organic Correlation of *Ficaria ranunculoides* during the Flowering Season" (27) arose from this bibliographical work and the erroneous manner in which he found Verschaffelt and MacLeod dealing with correlation. A further result of this work was that his confidence in the generality of the Mendelian hypotheses was much shaken. He found that Mendel's views were not consonant with the results formulated in a number of papers he had been led to abstract, and that the definite categories used by some Mendelian writers did not correspond to really well-defined classes in the characters themselves. It was a certain looseness of logic, a want of clear definition and scale, an absence of any insight into how far the numbers reached really prove what they are stated to prove, that moved Weldon when he came to deal with Mendelian work. And his attitude has been largely justified. The simplicity of Mendel's Mendelism has been gradually replaced by a complexity as great as that of any description hitherto suggested of hereditary relationships. This complexity allows of far greater elasticity in the deduction of statistical ratios, but the man in the street can no longer express a judgment upon whether the theory really accounts for the facts, and the actual statistical testing of the numbers obtained, as well as the logical development of the theory, will soon be feasible only to mathematical power of a high order. The old categories are, as Weldon indicated, being found insufficient, narrower classifications are being taken, and irregular dominance, imperfect recessiveness, the correlation of attributes, the latency of ancestral characters, and more complex determinantal theories are becoming the order of the day. If Weldon's papers " Mendel's Laws of Alternative Inheritance in Peas " (28), " On the Ambiguity of Mendel's Categories " (29), and " Mr Bateson's Revisions of Mendel's Theory of Heredity " (30), be read with a due regard to the dates of their appearance, it will be seen that they served, and that they continue to serve, a very useful purpose : they enforce the need for more cautious statement, for more careful classification, and for greater acquaintance with the nature of the inferences which are logically, i.e. mathematically, justifiable on the basis of given statistical data. The need will become the more urgent if the complexity of Mendelian formulae increases at the present rate.

To those who accept the biometric standpoint, that in the main evolution has not taken place by leaps, but by continuous selection of the favourable variation from the distribution of the offspring round the ancestrally fixed type, each selection modifying *pro rata* that type, there must be a manifest want in Mendelian theories of inheritance. Reproduction from this standpoint can only shake the kaleidoscope of existing alternatives; it can bring nothing new into the field. To complete a Mendelian theory we must apparently associate it for the purposes of evolution with some hypothesis of " mutations." The chief upholder of such an hypothesis has been de Vries, and Weldon's article on " Professor de Vries on the Origin of Species " (32) was the outcome of his consideration of this matter. During the years 1902 to 1903 an elaborate attempt was made to grow the numerous sub-races of *Draba verna*, with the idea that they might throw light on mutations. The project failed, largely owing to difficulties in the artificial cultivation of some of the species. But for a time all other interests paled before *Draba verna*.

"Where are you going at Easter? Stone wall country is very good, and if you find a place with delightful old stone villages and pretty churches, *Draba verna* will be there ! Come into this region, with the bike, and learn to know and love the dear Dog !*   Also explain to me how without thrashing to teach that same animal that lambs are not made to be eaten by puppies. There must be a way.   I have taught him to walk at heel past the most tempting of other dogs, and even past chickens, and I have not yet beaten him at all.   Cows and sheep will I suppose make one or two beatings necessary ? "   [Oxford, 7/2/02.]

And a little later comes a letter which shall be cited because it may induce another to take up a form of biometric work, which must some day be pushed to a successful issue ; in the fifteen years of letters there are many problems like this of *Draba verna*, which are discussed month after month with specimens, drawings, and tables, some merely schemed, but in surprising detail, others reaching the experimental stage, some in part solved, others but records of failure, one and all suggestive.

" *Draba verna*, or its earliest race is in full flower.   I have four model types from a certain wall.

Now can you and Mrs Pearson give us the week end, so that your eyes may see the glory of this plant ?

If you can turn up on Friday (I finish lecturing at 6) we can go for a tramp on Saturday, and see *Draba* at home on its walls.   A gentle 7 or 8 miles all told, in a decently pretty country, with a variable plant in the middle and a really Perfect Dog all the way makes a very good combination ;—only bring some knickerbockers, because Oxford Clay goes over one's ankles in places just now.

We can bring home our spoil, and discuss the very difficult question of descriptive units.

I think it a good and important thing to try.   All the problems of treating mixed races come in ; and above all I am curious to see what comes of statistical treatment applied to characters which have been chosen by "naturalists."   They all say we choose anything which is easy to measure, and neglect the real points of "biological" importance ; and there is a little truth in this reproach.

For *Draba* we want units of "habit," of shape and colour of leaves, of hairiness, of shape and colour of petals, sepals and fruit.   We want to treat leaves which are very distinctly differentiated according to their relative time of appearance, and I think having tabulated all these characters, we want to break up the plants on a wall which you shall see on Saturday into about four races.

Do come if you possibly can.   I saw one plant yesterday with all its seed capsules ripe and open ; so that the first lot of little races will very soon be out of flower.

* The great Borzoi Sandro, henceforward to be a marked feature of the Weldon household, at home and away from home.   Sandro pursuing sheep over the Yorkshire moors, Sandro pursuing game in the Buckinghamshire beechwoods, Weldon pursuing Sandro with every tone of affectionate persuasion, on the track the stacked cycles and the co-editor pursuing the deserted biometric problems in solitude, Weldon returning with the unchastised dog, after any interval of from 10 to 40 minutes, the chase being fully completed, the apologies for the Borzoi, his sustentation on chocolate and the human need for cigarettes, the return to the cycles, to the experiment that was to be crucial, to the colour and the sunset, these are all memories, the like of which others will have shared, which helped to form the atmosphere about the man.   Sandro achieved his purpose, he kept his master out in the air—such wolf hounds can follow a cycle for miles—and to exercise him was held up as a moral duty.   But his limited intelligence led to his own disablement and he had to become a partaker only in biometric " at homes." For two years, however, he was a great feature of our joint expeditions.

As *Nebensachen*, there are the mice and I very much want you to group the snails* in your own way—to see how far your grouping will bring them into a better form for the curve !

But *Draba verna* ought to be an example of the whole bearing of statistical methods upon systematic problems. I think it is rather important to begin this spring by collecting material for an analysis of the races round here, and a comparison of these with the types recognised by systematists. Learning these and their variability in this neighbourhood this spring, we can learn cotyledon characters and the characters of the first formed leaves in the Autumn—basing upon these a first set of hereditary correlations.

Again, the statement that each of Jordan's species can be recognised *at any stage in the life-history of the plant* makes it necessary to work correlations between cotyledon characters, autumn leaf characters, and spring characters. All this is work for lots of folk, and it is most important to get it properly planned now. Therefore, and for lots of other reasons try to come on Friday !

The Dog shall be washed for you !" [Oxford, 22/2/02.]

The reference in this letter to the mice indicates that that great piece of experimental work in heredity was now started. A study of the work of von Guaita had convinced Weldon early in 1901 that the cross between the European albino mouse and the Japanese waltzing mouse was not one which admitted of simple Mendelian description. In May, 1901, his letters contain inquiries as to Japanese mice dealers. During the summer and autumn the collection of Japanese mice was in progress. These mice were to be bred to test the purity of the stock ; during December about forty does had litters, and pure breeding went on until the autumn of 1902, when hybridisation commenced. The work on these mice was for two years entrusted to Mr A. D. Darbishire, but the whole plan of the experiments, the preparation of the correlation tables, and the elaborate calculations were in the main due to Weldon. On Mr Darbishire's leaving Oxford, Weldon again resumed personal control of the actual breeding arrangements, and from second hybrid matings carried on the work to the sixth hybrids' offspring. The work was nearing completion at his death, and through the energy of Mr Frank Sherlock the skins of the 600 pedigree mice forming the stud at that time have been dressed and added to those of the earlier generations. The reduction and publication of this material will, it is hoped, be not long delayed (33). Weldon had this work much at heart, and his letters during 1904 and 1905 give many indications of the points he considered demonstrated. The experimental part of the work would have been nearly completed had not his whole thought and energy been directed from November, 1905, into another channel.

From 1901 it is harder for the present writer to give a detailed account of Weldon's life, the co-editorship of *Biometrika* and common work brought them so continually into contact. In the early part of the summer there had been a hurried visit to Gremsmühlen for young *Clausilia* ; Weldon on his return visited his co-editor at Througham in Gloucestershire bringing his Brunsviga†, and there

* *H. nemoralis* and *H. hortensis* of which many hundreds had now been collected from various parts of England by Weldon and his helpers.

† The familiar mechanical calculator of the biometrician, the grinding sound of which (emphasised by the want of oiling in Weldon's case !) is the music which tells him how much his labours can be lightened.

was calculation and reduction of *Clausilia* data. Later there was a hurried visit to the Tegernsee and to Munich for opera. At Easter, 1902, there was a noble missionary effort (with the Brunsviga) to Parma; the missionary carried a memoir, which he had spent some weeks in rewriting in biometric form, but his efforts to show that a science of statistics exists were unavailing. In the summer *Biometrika* was edited from Bainbridge in Wensleydale, and accompanied by Sandro, the co-editors cycled to the churchyards of the Yorkshire dales, collecting material for their joint paper " On Assortative Mating in Man " (34). From Bainbridge the Weldons went to the British Association meeting at Belfast, where an evening lecture on Inheritance was given. At Christmas came one of the above-mentioned visits to Palermo to collect Sicilian snails. An event of this year (1902) was the publication of Mr Bateson's *Mendel's Principles of Heredity.* The origin of Weldon's first paper on Mendel has been described in this memoir; it was an expansion of a part of the promised bibliography for this Journal, and was written without any *arrière pensée* or knowledge of Mr Bateson's not then published experimental work. It is impossible for one who has been and again may be a combatant in this field to say more than that the tone of Mr Bateson's defence deeply pained Weldon, and rendered it difficult for a finely strung temperament to maintain—as it did maintain to the end—the impersonal tone of scientific controversy.

In the spring of 1903 Weldon was busy, as were the whole available members of the biometric school, in studying the influence of environment and of period of season on the variation and correlation of the floral parts of Lesser Celandine.

" Give my love to the Brethren who are cooperating in the matter of Celandines, and beseech them to make a better map of their country than the enclosed."   [Oxford, 23/2/03.]

Weldon threw his whole energy and love of minute exactitude into the task, and his letters are filled with an account of the almost daily changes in the type and variability of the Celandine flowers from his selected stations. The result of this enquiry was the collection of an immense amount of data showing that environment and period in the flowering season affected the flower characters to an extent comparable with the differences attributed to local races. The reduction of the material has gone on progressively, if intermittently since, and it is hoped that a memoir, which will be a sequel to that issued in *Biometrika**, may be published shortly (35). The wider standpoint of this second memoir will be chiefly due to Weldon's initiative and critical mind. At Easter of 1903 a series of mishaps prevented the common holiday, but this was more than compensated for by the summer vacation. The Weldons started with a sea trip to Marseilles and back. They then returned to Oxford, that work on the article *Crustacea* for the Cambridge Natural History might be carried on, and an eye kept on the mice. But a biometric camp was formed at Peppard on the Chilterns; here the " Consulting Editor " and one of the co-editors had established themselves, and the Weldons took a week-end cottage. The three Oxford members of the party arrived partly on cycles and partly on four

---

* Vol. II. pp. 145—164, *Cooperative Investigations on Plants.* II. Variation and Correlation in Lesser Celandine from Divers Localities.

feet, and were often met *en route* by the residents in the uplands, the numbers being swelled by the addition of biometricians from the London or Oxford schools. Hence arose a series of Friday "biometric teas," for the discussion of the week's work and plans for the next two days. Saturday and Sunday morning were given to steady calculating and reducing work, and much was got through. The data on assortative mating in man collected in the previous year were reduced and a joint paper sent to press; the immense amount of calculation and reduction involved in the mouse-paper was got through; a joint criticism of Johannsen's *Ueber Erblichheit in Populationen und in reinen Linien* was written by the co-editors under the title "Inheritance in *Phaseolus vulgaris*" (36); the Huxley Lecture was written with yeoman help from the Oxford contingent, and lastly, a joint study was made, at Weldon's suggestion, of the relationship between Mendelian formulae and the theory of ancestral heredity. It was shown that there was no essential antagonism between the two methods of approaching the subject, and the results were published ultimately as Part XII of the *Mathematical Contributions to the Theory of Evolution*, Weldon persistently declining to allow it to appear as a joint memoir, because he had taken no part in certain portions of the more complicated algebraic analysis. Christmas found two-thirds of the party reunited in Palermo, and Weldon on the snail quest. His letters thence to his co-editor teem with the freshness of the sky and the joy of open-air work :

"Out between five and six, in the dark, without any breakfast, sunrise up in the hills, a day's tramp on a piece of bread and a handful of olives, and home at seven, laden with snails. Then after dinner to clean the beasts. That is not work, and it makes one very fit, but one gets tired enough to sleep when the snails are cleaned !

The camera works all right, and I think there is a very marked correlation between the general character of the limestones and the character of the shells ; but developing in one's bedroom does not make for negatives which will "process" ! Also it involves heavy subsidies to such chambermaids as do not understand what new form of madness this particular foreigner has developed !

I have repeated all last year's collections, and have tried hard to get a series of forms, such as Kobelt describes, intermediate between the rounded and the flat keeled forms, but I cannot at present find these. They ought (according to him) to live in a certain wilderness of beautiful mountains twenty miles away. I have several times tramped without any result. I hope to try again. I feel sure something worth having will come out of these shells ; they illustrate local races and the general problem of what is a species splendidly. But the question of their markings comes in also ; and you, or Galton, or someone, will have to make a scale of patterns for me, I expect. They will be the most perfectly hopeless things to draw !

It is, of course, just conceivable that the intermediate, slightly keeled forms described in 1879 by Kobelt have been exterminated since his time ? He is very precise in his localities, and everywhere except in his transitional region I find *exactly* what he describes, but in this region I find so far only rounded forms.

The only difficulties about tramping in this country are the *carabinieri*. Every high road is patrolled by groups of two or three, so that even in a desolate place, so long as you keep to the road, you are rarely out of rifle shot ; but these men come and solemnly warn you that the people round are ruffians, who would cheerfully cut your throat for a soldo ; and if you simply grin, they make a great pretence of falling in behind you and guarding you....Now collecting snails with an armed guard becomes ridiculous after a time and there is no danger at all ; the men only want tips. When one gets off the road into the hills the goatherds and other ruffians are most friendly. They want to see one's camera, and one's knife, and of course they want half one's bread, but they never ask for tips, and my throat is still uncut.

We have so far had two wet days ; to-day, and one other.—We have had several inches to-day, and shall have some more ; but between we have had the most glorious sun. I look as if I washed in strong coffee every morning."   [Palermo, 31/12/1903.]

At the beginning of 1904 the work on the Brescia *Clausilia* was in progress, the mice were multiplying after their kind and Weldon's thoughts were turning more and more to a determinantal theory of inheritance, which should give simple Mendelism at one end of the range and blended inheritance at the other. Easter was spent in common, one editor at Rotherfield Greys and the other at Peppard, with the usual flow of suggestion on Weldon's part and the bi-weekly cyclings to Oxford to look after the mice. Now and then the fear would strike Weldon's friends that life was being lived at too fast a pace, but the constant intellectual and physical activity was so characteristic of the man that there was no means of calling halt, and to many when Weldon was most active he seemed most fit and well. The summer found the Pearsons twelve miles from Oxford, at Cogges, near Witney, and the Galtons twenty miles further, at Bibury ; there was much cycling too and fro, and the plan of a new book by Weldon on Inheritance was drafted, and some of the early chapters written. The vacation was broken by the visit to Cambridge—Weldon cycling in one day from Oxford—for the British Association. The Presidential Address in the Zoological Section was chiefly an attack upon biometric work and methods, and the discussion which followed culminated in the President dramatically holding aloft the volumes of this Journal as patent evidence of the folly of the school, and refusing the offer of a truce to this time-wasting controversy. The excitement of the meeting, as earlier contests at the Zoological and Linnean Societies, seemed to brace Weldon to greater intellectual activity and wider plans, but the torpedo boat was being run at full speed.

The book on Inheritance occupied most of the remainder of the year, and to aid it forward and help those of us who were not biologists to clearer notions, I suggested to Weldon a course of lectures in London to my own little group of biometric workers. The project grew, other departments of the College desired to attend, and ultimately the lectures were thrown open to all members of the University and even to the outside public. Weldon had a good audience of more than a hundred, and enjoyed the return to his old environment. But it may be doubted whether his vitality responded as quickly as of old to the additional stress ; there were special elements of difficulty, and I believe now that it would have been kinder and more helpful had we limited the audience to my own small body of sympathetic students.

"It will be a great pleasure to me to come and talk, and to feel that you cared to ask me; the lectures will do far more good to me than to anyone else ... and I owe U. C. L. a bigger effort than this anyhow."   [Oxford, 16/10/04.]

And again:

"It makes me more than ever glad I am coming back to Gower Street where there are live people to talk to! Surely thirty people* is a great many. Try talking for five years to an audience of from three to nine, and see how the thought of thirty will cheer you! And none of these excellent folk are sent by their tutors!" [Oxford, 7/11/04.]

The letters of Weldon to both Francis Galton and myself during the years 1904 and 1905 are full of inheritance work, the details of the great mice-breeding experiment, the statement and the solution, or it might be the suggested solution, of nuclear problems leading to determinantal theories of inheritance. Occasionally there would be a touch of conscience, and the drawings for the *Crustacea* would be pressed forward:

"I ought to give my whole time to the Cambridge Natural History for a while. They have been very good to me, and I have treated them more than a little badly. I am rather anxious to get them off my conscience." [Oxford, 15/2/05.]

But only the chapter on *Phyllopods* got completed, figures and all, and set up. Many figures were prepared for other parts; beautiful things, which gave Weldon not only scientific but artistic pleasure, he had made, but the text remains the veriest fragment. In the same way but little was absolutely completed of the article on *Heliozoa* for Lankester's *Natural History*. It was not Weldon's biometric friends that kept him from these tasks, it was solely his own intense keenness in the pursuit of new knowledge. It was occasionally with a feeling of great responsibility that the present writer would propound to him an unsolved problem with which he might himself be struggling. There was absolute certainty that if the problem was at all an exciting one, Weldon would leave his scent and follow the new trail with his whole keenness and at full speed. All else would be put on one side, and he could only be recalled to natural history or biometrics by an appeal to his conscience. Like Sandro, the chase must be completed before he returned to the humdrum trot behind a cycle on the highway.

The fascination of inheritance problems kept Weldon, however, for months at a time at the Heredity Book. At Easter, 1905, he went to Ferrara†, because that place had a university, and as such must have a library, where work could be done. The contents of the library were perfectly mediaeval, a characteristic appropriate in the castle, but hardly helpful in heredity‡. Still, portions of the manuscript came to England for comment and criticism, and we were hopeful that the end of the year would see the book completed.

* The number I already knew would certainly attend.

† "The town is worth a lot, and the fields are full of a little speedwell, which varies most delightfully. I have so far resisted the temptation to chuck the wretched book and tabulate the variations of its flowers, and I hope I shall do to the end. But it is a temptation......I feel out of the world, an absolute blank, with only a slight interest in newts' tails and an even slighter in a statue of Savonarola which looks at me all day through the window." [Ferrara, 3/4/05.]

‡ From Ferrara came back if not the speedwells, masses of silkworms' eggs of different local races, but providentially they failed to hatch out in sufficient numbers owing to the May frosts and no new scent took Weldon off the book and the mice during the summer.

It must not be thought for a moment that Weldon was desultory in his work. As E. R. Lankester says in a letter to the writer: "His *absolute thoroughness* and unstinting devotion to any work he took up were leading features in his character." He pursued science, however, for sheer love of it, and he would have continued to do so had he been Alexander Selkirk on the island with no opportunity for publication and nobody to communicate his results to. He never slackened in the total energy he gave to scientific work, but having satisfied *himself* in one quest, he did not stay to fill in the page for others to read; his keen eye found a new problem where the ordinary man saw a cow-pasture, or a dusty hedgerow, and he started again with unremitted ardour to what had for himself the greater interest. The publication of his researches will show that it is not we who are the losers, because he went forward, regardless of publication and finality of form. The true function of such a man is not to write text-books or publish treatises, it lies in directing and inspiring a school, which will be trained by completing the work and carrying out the suggestions of its master. The curse of the English educational system is that it leaves such men to solitude, and throngs the chambers of those who cram all nature into the limits of the examination room.

In the summer the present writer was at East Ilsley, some seventeen miles from Oxford, and there was cycling out several times a week; the writer's chief work was on other than biometric lines and broken by other claims on his time, but there was steady joint work on the determinantal theory of inheritance as outlined by Weldon, and it is hoped that it is sufficiently advanced to be completed and published (37). Weldon had in August, 1905, given to the Summer Meeting of the University Extension in Oxford a lecture on *Inheritance in Animals and Plants* (38), and this had taken up some of his energy during the summer vacation. On the whole, however, he worked persistently at the Inheritance Book. It is too early yet to say definitely how far it can be considered ready for press, but a considerable number of chapters are completely ready, and there are drafts for several others. We can only hope that this, the work he was in many respects best fitted for both by direct experiment and by study of the labours of others, will be issued in his name and show the full measure of his activities during these last few years.

It cannot be denied that those who were often with Weldon during the last two years were occasionally anxious—the pace had been too great—but at no time had one definitely realised that there was an immediate anxiety. His intellectual activity was never apparently diminished, and his long cycling rides were maintained to the end. It was an occasional, but never long persistent, lack of the old joyousness in life which was noticeable. At East Ilsley he was full of keenness over his photographic work; he enjoyed an antiquarian investigation into the probable final *locus* of the bones of St Birinus with a view to testing a local legend; we examined carefully a human skeleton dug up from under a sheepfold, the authorities having determined that no inquest was needful, the bones being those of an old man who died "hundreds of years ago." "And you think?" said

To the Editor of Biometrika

Sir, — In the last number of your most instructive Journal you publish an account of the way in which Smiles are collected by the energetic naturalists of Brunt. It may bring in harvest to those of your numerous readers to brush up the knowledge neglected by their account, and to know something of the way in which Smiles grow up upon their own collection.

To you, Sir, turn very well, a Smile is unlike most of the animals found in the neighbourhood of Brunt, in its shape. Most of the animals in the Brunt collection, whether Bats, or Butterflies, or Caterpillars, or others, can be divided into halves, which are curiously alike. If you take a thing like the drawing in the margin, which is meant to be like half a crayfish, and put the straight edge of the drawing against a looking glass, the reflection of this drawing in the glass will be very like the other half of the crayfish. See the legs of a crayfish are in pairs, so that there are right and left legs, as curiously like each other as the Editor's Right Hands are like each other.

But except the two halves of its brain, and its eyes, and one or two others again on its head, the parts of a grown up Smile are not arranged in pairs in this way. When a Smile is quite young, inside its egg-shell, you can recognise a right and a left half of it, just as you can in a crayfish. But as it grows up, its left hand

Plate V.

Now a snail never has any legs; but it has a number of paired organs, right and left, in the hinder part of its body, when it is young. As it grows, it twists to the hind end of its body towards as I have described; and while it is doing this, he as one which lay on its left side, its left kidney, the left half of its heart, and other things, disappear. So that it only has one kidney, and the piece of its heart which lays on its right side when it was young and ~~constant~~ had not got ~~got~~ got a large hump on its left side.

Now this left-hand hump, which is produced by the growth of the left side, gets twisted like a screw, and the shell is formed on it. So that when you see this kind of thing when you look at a full-grown snail you could see this kind of thing:

This is the hump twisted out of the left side of the snail and screwed up

over here are the young things right hand side of body towards you

This is the hind of your snail.

There are the horns of the snail, in front, the place where its left side lays by its side up, so

(that the right & left sides are alike.)

Nearly all snails screw their hump, after they have formed it, the same way round. They screw it so that if you stand on the top of the hump, and imagine yourself walking down the twisted hump towards the place where it joins the rest of

snail, you will always have turned to the right. But some snails screw their humps the other way, so that in going from the top to the bottom of their hump you would always have to turn to the left. Among the snails found near Captain Beechy, very one has a right handed hump, except Clanilia, which has a left hand hump. Only every now and then a snail of a sort which has usually a right-handed hump makes a left handed hump for itself. Nobody knows what the children of such wrongly twisted snails are like; whether they are like their parents, or whether they return to more usual ways of twisting their hump. And nobody knows how a snail is better off for going through this preap of enlarging one side and twisting it; or of having no hump of all; so that certainly no one knows whether, if a snail must have a twisted hump, it is better to have a right-handed or a left-handed twist.

Because no one knows these things, some very great people, who dislike Biochronicles, say that it does not matter to a snail which way its hump is twisted. They say that it is absurd to suggest that such a trifle, the meaning of which they cannot understand, can have any effect upon a snail. I venture to hope, Sir, that as Biochronicles purspgge you will take an ever-increasing interest in snails, but I hope you will never be led to suppose in your influential columns the opinion that nothing matters to a snail unless you can yourself understand it.

I am, Sir, Your obedient servant, A.A.Wheldle

Weldon on our homeward way. " Having no anatomical training I think they are those of ——." " A young woman, who has not been buried so very long," he interrupted, with a responsive twinkle in his eye. " Let us have a smoke and consider the scientific education of the English medical profession." His sense of humour was always keen, whether with word or pencil, and it remained with him to the end. The joy of life which in the early days led him to dance and sing on the completion of a heavy bit of work, made him in later manhood ripple over with quiet humour in talk and letter when problems were going well.

Thus to Francis Galton :

" I enclose the best I can do with one of the negatives you were kind enough to let me make. Please forgive me for caricaturing you in this way.—You know enough about the lower forms of man to know that respect and affection show themselves in strange ways :—look upon this as one of them and pardon it." [Oxford, 27/7/05.]

Nor did he spare a quiet joke at a friend :

" Your work on dams has filled the Italian papers with horror. They say you threaten the safety of all existing dams, *however long they have stood.*" [Ferrara, 7/4/05.]

In November, 1905, Weldon was unfortunately taken off from the work on his inheritance book by the presentation to the Royal Society of a paper by Captain C. C. Hurst : *On the Inheritance of Coat-Colour in Horses.* He had had no proper summer holiday, but he threw himself nine hours a day into the study of *The General Studbook**.

" I can do nothing else until I have found out what it means....The question between Mendel and Galton's theory of Reversion ought to be answered out of these. Thank God, I have not finished that book. There must be a chapter on Race Horses ! "

Weldon felt himself in a difficult position; as Chairman of the Zoological Committee, he had at once directed the printing of Hurst's paper. But the subject being one in which he personally was keenly interested, he had immediately attacked the original material and to his surprise came to views definitely opposite to those of Hurst. He felt bound to report this result at once to the Society, and he did so on December 7, when the original paper was read. His results were provisional, as could only be the case considering the short period of preparation that had been possible. He promised to communicate a note to the Society involving more details of his inquiry. This was done on January 18, 1906 in a "Note on the Offspring of Thoroughbred Chestnut Mares " (39).

---

* I cannot resist citing a last illustration of Weldon's humour: ''What volumes of Weatherby have you? I have found in Bodley 17—20. To show you what Bodley is, I looked in the catalogue vainly under: *Weatherby* (found here and not under Racing, *Racing Calendar*), *Jockey Club* (found here pamphlets about the J. C. but *not* its own publications), *Horses, Race Horses, Racing, Studbooks* (found here only Clydesdale Studbook, Pigeon Studbooks, and Dog Studbooks), *Turf, Sport, Race*, all suggested by assistants in the Library. For a whole day I raged, and came back despairing. Next day I raged worse, and captured a man who knew something. He smiled and said: ' Oh, Yes, The General Studbook is entered under *General* of course.' I said, 'Why not under *The*?' and he thought that unseemly ! "

"The object of the present note is partly to fulfil my promise and partly to call attention to certain facts which must be considered in the attempt to apply any Mendelian formula whatever to the inheritance of coat-colour in race-horses."

It is impossible at present to say more on this point, for the whole subject is likely to be matter for further controversy. Even one authenticated case of a non-chestnut offspring to chestnut parents is sufficient to upset the theory of the 'pure gamete,' but if studbooks are to be taken as providing the data, the whole question must turn on whether one in sixty of the entries of the offspring of chestnut parents can be reasonably considered as a misprint or an error in record.

Here it can only be said that Weldon took up the subject with his usual vigour and thoroughness. But he was overworked and overwrought and a holiday was absolutely needful. He went to Rome, but the volumes of the Studbook went with him:

"Will you think me a brute, if I take the Studbook to Rome? I really want a holiday, but I cannot leave this thing unsettled."

And then from Rome:

"I think it will be worth while to deal for once with a whole population, not with a small random sample. Only I could find it in my heart to wish one need not do it in Rome! To sit here eight hours a day or so, doing mere clerk's work, seems rather waste of life?"

And again:

"I have really been working too hard to write, or to do anything else. I have seen *nothing* of Rome....I want to know what these horses will lead to, but it would not interest me at all to know that my paper on them would or would not be printed. More important is the enormous time these horses will take. It seems clear that one ought to carry these arrays back to another generation of ancestors—and that means a very long job. I wish I had a pupil! A mere clerk would be no good, but a pupil, such as one had in good old Gower Street, would help with the drudgery, and then he might stick his name all over the paper, if he liked."

[February, 1906.]

The letters are filled with Studbook detail till Easter, there is hardly a reference to anything else. Re-reading them now one sees how this drudgery with no proper holiday told on Weldon. Hundreds of pedigrees were formed and a vast amount of material reduced. At Easter the Weldons went to the little inn at Woolstone, at the foot of the White Horse Hill, and his co-editor came down later to Longcot, a mile away, for the joint vacation. Weldon was still hard at work on the Studbooks, but he was intellectually as keenly active as of old; he was planning the lines of his big memoir on coat-colour in horses (40) and showing how they illustrated the points he had already found in the mice. He was photographing the White Horse, and rubbing mediaeval idlers' scrawlings on the church pillars. He projected the despoiling of a barrow, and planned future work and rides.

On Sunday, April 8, he rode into Oxford to develop photographs, and the present writer rode some miles of the way with him; the joint ride terminated with the smoke by the roadside and Weldon's propounding the problem which was to be brought solved for him on Tuesday. On Tuesday I found him in bed, with what appeared to be an attack of influenza. He had expressed himself tired after his ride on Sunday, an almost unique admission. But on Monday he went a long walk over the Downs, getting home late. He came down to breakfast on Tuesday but had to return to bed. In the afternoon when I came he insisted on smoking and wanted the solution of the problem, saying he was better. I begged him, as one still closer did, to stay in bed on the morrow and give up a projected journey to Town. But there was a dentist to be seen, preparations for a visit to the M.B.A. Laboratory at Lowestoft to be made, and a wonderful picture-gallery to be visited to free him from the atmosphere of the Studbooks. His will was indomitable ; he went up to Town and went to the pictures on Wednesday, he went to the dentist on Thursday, but from the dentist's chair he had to be taken to a doctor's, and thence to a nursing home. The summoning telegram reached his wife on the same afternoon, and he died of pneumonia on Good Friday, April 13. So passed away, shall I say not unfitly—for it was without any long disabling illness and in full intellectual vigour—a man of unusual personality, one of the most inspiring and loveable of teachers, the least self-regarding and the most helpful of friends, and the most generous of opponents.

As for his life, I think it was to him what he would have wished it. There were moments of discouragement and depression, he felt occasionally a want of sympathy for his life-work in some of his former colleagues, and while he was born to be the centre of an enthusiastic school, he found at times somewhat scanty material for its maintenance in pleasure-loving Oxford. But every stone he lifted from the way became gold in his hands; each problem he touched became a joy which absorbed his whole being. The artist in his nature was so intense that he found keen pleasure in most men and in all things. Only meanness or superficiality fired him, and then, considering how the world is built, sometimes to almost an excess of wrath. But he had no personal hate ; he could make the graceful amend, and had he ever a foe, that foe, I veritably believe, could have won Weldon's heart in the smoking of a cigarette.

If we pass from himself to those whose fortune brought them in close contact with him—to his friends and pupils—their loss can only be outlined, it is too intimate and personal for full expression. There was a transition from respect to reverence, a growth from affection to love ; to such a tenderness as some bear for a more delicate spiritual nature, to even such feeling as the Sikh is reputed to hold for the white man's child in his charge.

And lastly as to science, what will his place be ? The time to judge is not yet. Much of his work has still to be published, and this is not the occasion to indicate what biometry has already achieved. The movement he aided in starting is but in its infancy. It has to fight not for this theory or that, but for a new method and a greater standard of logical exactness in the science of life. To those who condemn it out of hand, or emphasise its slightest slip, we can boldly reply, You simply cannot judge, for you have not the requisite knowledge. To the biometrician, Weldon will remain as the first biologist who, able to make his name by following the old tracks, chose to strike out a new path—and one which carried him far away from his earlier colleagues. It is scarcely to be wondered at if those he joined should wish to see some monument to his memory; for he fell, the volume of life exhausted, fighting for the new learning.

Is what he gave science small? That depends on how it is measured. He was by nature a poet, and these give the best to science, for they give ideas. They follow no men, but give that which another generation may study from and be inspired by. He was the enthusiast, but the enthusiasm was that of the study, trained to its task; and when the time comes that we shall know, or that those who come after us shall know, whether Darwinism is the basal rule of life or merely a golden dream which has led us onwards to greater intellectual insight, then the knowledge, so biometricians have held and still hold, will be won by those actuarial methods which he first applied to the selection of living forms. If there be aught else to be said, let another say it.

> Step to a tune, square chests, erect each head,
>   'Ware the beholders!
> This is our master, famous, calm and dead,
>   Borne on our shoulders.

Description of Plates.

Plate I. W. F. R. Weldon.

Plate II. Raphael Weldon, aged 10.

Plate III.　(*a*)　Rapid pencil caricature by W. F. R. W. "Apparition: Le Café Orleans."

　　　　　(*b*)　Sample of Illustration to letters. Description of bands of *H. hortensis* in letter to a lady collector. "Has it occurred to you that a lady of artistic ability, and so enlightened that she likes snails, would have great joy and do great service by drawing them? There is a good inexorable severity about their lines which one would enjoy, I should think, if it were not so unattainable (to me!) on paper. And it would be nearly as good fun as real engraving to get all their lights and shadows put in with curved lines which also indicate the growth lines on the shell? Think how Bewick liked it."

Plate IV. A "crabbery" at Plymouth.

Plate V. Contribution to a manuscript magazine run by a youthful friend.

LIST OF MEMOIRS, ETC., BY W. F. R. WELDON.

(1) Note on the early Development of *Lacerta muralis*. *Q. Jour. Mic. Sci.* Vol. XXIII, pp. 134—144, 1883.

(2) On the Head-Kidney of *Bdellostoma*, with a suggestion as to the Origin of the Suprarenal Bodies. *Q. Jour. Mic. Sci.* Vol. XXIV, pp. 171—182, 1884.

(3) On the Suprarenal Bodies of Vertebrates. *Q. Jour. Mic. Sci.* Vol. XXV, pp. 137—150, 1885.

(4) On some points in the Anatomy of *Phoenicopterus* and its Allies. *Proc. Zool. Soc. Lond.* 1883, pp. 638–652, 1883.

(5) Note on the Placentation of *Tetraceros quadricornis*. *Proc. Zool. Soc. Lond.* 1884, pp. 2—6, 1884.

(6) Notes on *Callithrix gigot*. *Proc. Zool. Soc. Lond.* 1884, pp. 6—9, 1884.

(7) On *Dinophilus gigas*. *Q. Jour. Mic. Sci.* Vol. XXVII, pp. 109—121, 1886.

(8) *Haplodiscus piger* ; a new Pelagic organism from the Bahamas. *Q. Jour. Mic. Sci.* Vol. XXIX, pp. 1—8, 1888.

(9) Preliminary Note on a *Balanoglossus* Larva from the Bahamas. *R. S. Proc.* Vol. XLII, pp. 146—150, 1887.

(10) The Coelom and Nephridia of *Palaemon serratus*. *Journal Marine Biol. Assoc.* Vol. I, pp. 162—168, 1889.

(10) *bis* Note on the Function of the Spines of the Crustacean *Zooea*. *Journal Marine Biol. Assoc.* Vol. I, pp. 169—170, 1889.

(11) The Renal Organs of certain Decapod Crustacea. *Q. Jour. Mic. Sci.* Vol. XXXII, pp. 279—291, 1891.

(12) The Formation of the Germ Layers in *Crangon vulgaris*. *Q. Jour. Mic. Sci.* Vol. XXXIII, pp. 343—363, 1892.

(13) The Variations occurring in certain Decapod Crustacea. I. *Crangon vulgaris*. *R. S. Proc.* Vol. XLVII, pp. 445—453, 1890.

(14) Certain correlated Variations in *Crangon vulgaris*. *R. S. Proc.* Vol. LI, pp. 2—21, 1892.

(15) On certain correlated Variations in *Carcinus moenas*. *R. S. Proc.* Vol. LIV, pp. 318—329, 1893.

(16) [On Variation in the Herring. Unpublished measurements and reductions presented to the Evolution Committee.]

(17) Attempt to measure the Death-rate due to the Selective Destruction of *Carcinus moenas* with respect to a Particular Dimension. Report of the Committee...for conducting Statistical Inquiries into the Measurable Characteristics of Plants and Animals. *R. S. Proc.* Vol. LVII, pp. 360—379, 1895.

(18) Remarks on Variation in Animals and Plants. *R. S. Proc.* Vol. LVII, pp. 379—382, 1895.

(19) [Report to the Evolution Committee on the Growth of *Carcinus moenas* at successive moults. 1897. Unpublished.]

(20) Presidential Address to the Zoological Section of the British Association. *B. A. Transactions*, Bristol, 1898, pp. 887—902.

(21) [Researches on Pedigree Moths, 1899—1901. Unpublished.]

(22) Cooperative Investigations on Plants. I. On Inheritance in the Shirley Poppy. *Biometrika*, Vol. II, pp. 56—100, 1902. [A joint paper with others.]

(23) A First Study of Natural Selection in *Clausilia laminata* (Montagu). *Biometrika*, Vol. I, pp. 109—124, 1901.

(24) Note on a Race of *Clausilia itala* (von Martens). *Biometrika*, Vol. III, pp. 299—307, 1903.

(25) The Scope of *Biometrika*. Editorial. *Biometrika*, Vol. I, pp. 1, 2, 1901.

(26) [Critical Bibliography of Memoirs on Inheritance. Unpublished.]

(27) Change in Organic Correlation of *Ficaria ranunculoides* during the Flowering Season. *Biometrika*, Vol. I, pp. 125—8, 1901.

(28) Mendel's Laws of Alternative Inheritance in Peas. *Biometrika*, Vol. I, pp. 228—254, 1902.

(29) On the Ambiguity of Mendel's Categories. *Biometrika*, Vol. II, pp. 44—55, 1902.

(30) Mr Bateson's Revisions of Mendel's Theory of Heredity. *Biometrika*, Vol. II, pp. 286—298, 1903.

(30) *bis* Mendelism and Mice. *Nature*, Vol. LXVII, pp. 512, 610, Vol. LXVIII, p. 34, 1903.

(31) Albinism in Sicily and Mendel's Laws. *Biometrika*, Vol. III, pp. 107—109, 1904.

(32) Professor de Vries on the Origin of Species. *Biometrika*, Vol. I, pp. 365—374, 1902.

(33) [On the Results of Crossing Japanese Waltzing with Albino Mice. Unpublished.]

(34) On Assortative Mating in Man. *Biometrika*, Vol. II, pp. 481—498. A joint memoir, 1903.

(35) [Measurements and observations on Lesser Celandine. Unpublished.]

(36) Inheritance in *Phaseolus vulgaris*. *Biometrika*, Vol. II, pp. 499—503.   Joint review, 1903.

(37) [A Determinantal Theory of Inheritance. Unpublished.]

(38) Inheritance in Animals and Plants. *Lectures on the Method of Science*. Edited by T. B. Strong, Oxford, 1906.

(39) Note on the Offspring of Thoroughbred Chestnut Mares. *R. S. Proc.* Vol. 77B, pp. 394—398, 1906.

(40) [Material for an extensive memoir on the Inheritance of Coat-colour in Thoroughbred Horses. Unpublished.]

(41) Article on *Crustacea* for the *Cambridge Natural History*—fragmentary, except for a chapter on the *Phyllopods* already set up.

(42) [A Treatise on Inheritance, largely completed.]

(43) A portion of an account of the *Heliozoa* for the *Oxford Natural History*.

(44) Account of Kölliker's scientific work. *Nature*, Vol. LVIII, pp. 1—4, 1898.

(44) *bis* Dreyer's Peneroplis, eine Studie zur biologischen Morphologie und zur Speciesfrage. (A review.) *Nature*, Vol. LIX, pp. 364—5, 1899.

(45) Account of Huxley's scientific work for the Supplement to the *Dictionary of National Biography*, 1900.

(46) Article on *Variation* in the " Times" Supplement to the *Encyclopaedia Britannica*.